Literature-Based Science Activities

by
Audrey Brainard
and
Denise H. Wrubel

SCHOLASTIC
PROFESSIONAL BOOKS

New York • Toronto • London • Auckland • Sydney

To all the teachers - whose dedication, imagination, and motivation inspire us all.

To all the students - may they enjoy and learn from exploring books and science experiments.

To Jackie - for her enthusiasm and optimism.

To our husbands - for all their encouragement and understanding.

Design by Nancy Metcalf
Production by Intergraphics
Cover design by Vincent Ceci
Cover illustration by Therese Anderko
Illustrations by Michele Fridkin

ISBN 0-590-49200-4
Copyright © 1993 by Scholastic Inc. All rights reserved.

Printed in the U.S.A.

Contents

Continued

About This Book

We believe that science, children, and books are a natural combination. The following pages present ideas to help you link science concepts with great children's books.

We selected 20 science topics frequently taught in the early grades. For each topic we highlighted two books that can be used to introduce, extend, or enrich the science content. Additional titles are suggested under **Literature Tie-Ins**. Whether a book is fiction (F), non-fiction (NF), or poetry (P) is indicated in parenthesis with the publishing information. To help relieve apprehension, we've included a **Background** section for each topic. This summarizes and briefly explains the core concepts that the children need to know.

Within each topic each section begins with **Activities** correlated to the highlighted book. To facilitate interdisciplinary activities we've included ideas for **Curriculum Crossovers**. Suggestions for having students explore the world outside the classroom and for bringing the outside world into the classroom are given under the heading **Take a Trip/Invite a Guest**. For added enrichment, reproducible worksheets accompany the books.

Sharing Books with Children

Sharing books with children is one of the most valuable and long-lasting experiences we can have with them. It exposes them to correct usage of the English language, expands their vocabulary, and opens up vistas for their imagination.

The most common method of book sharing is to sit with the students in a circle and hold the book so that they can see the pictures while you read the text. This method should not be abandoned, but some of the following alternatives could add variety to book sharing time.

Read the book without showing the cover or the pictures. Stop at appropriate times and have the children draw what they imagine the scene to be. After you finish the story, allow time for students to share their pictures before you show them the book illustrations.

If Big Books are unavailable, consider obtaining two copies of a book. While you read from one copy, an assistant holds the other copy open for the class to see the pictures. By standing slightly behind you the assistant can easily see when to turn the page.

Rather than make story characters out of flannel, consider using non-iron-on interfacing found in fabric stores. The interfacing easily adheres to the flannel board with a firm touch of the hand. To make the characters, lay the interfacing on top of the pictures and trace the desired outline. Remove the interfacing and color it with crayon or felt marker. As you read or tell the story, you can add characters to the board. After the first reading children can assist in placing the appropriate cutouts on the board. Students may also retell the story using the book characters.

Enjoy the magical combination of children, science, and books!

Shadows

- *BEAR SHADOW* by Frank Asch.
- *NOTHING STICKS LIKE A SHADOW* by Ann Tompert.

Background:

1. Some materials allow light to pass through them, and some materials do not.

2. Transparent materials, such as clear glass, allow nearly all the light to pass through. If you look through a piece of clear glass you see a clear image.

3. Translucent materials, such as wax paper, filter the light that passes through them. If you look through a piece of wax paper, images appear fuzzy.

4. Opaque materials block all of the light. You cannot see anything through opaque materials.

5. In order to create a shadow you need a source of light, an opaque object (something that blocks light), and somewhere for the shadow to fall.

Literature Tie-Ins:

- **HENRY AND THE DRAGON** by Eileen Christelow (Clarion Books, 1984, F)

- **ME AND MY SHADOW** by Arthur Dorros (Scholastic Hardcover, 1990, NF)

- **MY SHADOW** by Robert Louis Stevenson (G.P. Putnam's Sons, 1990, F)

- **SHADOWS ARE ALL ABOUT** by Ann Witford Paul (Scholastic, 1992, NF)

- **MR. WINK AND HIS SHADOW, NED** by Dick Gackenbach (Harper & Row Publishers, 1983, F)

- **I HAVE A FRIEND** by Keiko Narahushi (Margaret K. McElderry Books, 1987, F)

 ## Bear Shadow

by Frank Asch (Simon and Schuster, 1985, F).

Bear's shadow gets in his way when he is trying to catch a fish. What can Bear do to get rid of his shadow?

Activities:

1. Each child will need to make a cardboard or stiff paper figure that will be cut out and used to cast a shadow following the directions below. Any shape figure will do; e.g. gingerbread people, teddy bears, etc..

Each child will also need a pencil, a thread spool, and adhesive tape. Using tape, they attach the figure to the eraser end of the pencil. Then they support the figure by inserting the pointed end of the pencil into the thread spool.

Ask the children what they think will happen when the figure is placed in the sun. Record their guesses. Take the children outside on a sunny morning. Have them place large sheets of paper on the ground. Ask the children to place their supported figure on their paper. Using crayon, markers, or chalk the children then trace the shadows onto the paper. Repeat the activity at noon and 2 p.m. Ask the children to describe how the figure's shadow changed during the day.

2. Ask the children what they can do to "get rid" of the shadow made by the figure. They can

try using assorted materials (construction paper, a clear material such as a transparency or clear plastic wrap, wax paper, or tracing paper) to block the shadow. Ask the children to describe what happens when they try to block the light with each of the materials.

3. Have the children sort the materials they used to try to block the shadow into three groups: materials that block all of the light, materials that block some of the light, and materials that do not block any of the light.

4. Children can explore the relationship between the position of the light source and the shape of the shadow by attaching a 45cm piece of string to the neck of a flashlight. Give the children the following materials and instructions:

   a. Place a piece of paper on your work surface.

   b. Use a piece of clay to attach a short pencil to the center of the piece of paper.

   c. With a thumb, hold the loose end of the string on the work surface beside the pencil.

   d. Hold the flashlight parallel to the work surface, yarn stretched tightly. Be sure the flashlight is on. Slowly move the flashlight in an arc over the pencil from one side to the other. Observe the changes in the pencil shadow. Use crayons to record the length and location of the shadows. Describe the changes you observe.

 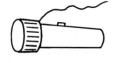

5. Early humans used shadows to tell the time of day. Shadows appear long in the morning, shrink toward midday, and then grow longer until dusk. Ask children to record the location and length of the shadow formed by the flagpole at 9 a.m., noon, and 2 p.m. Students can also measure the shadow cast by their pencils in the thread spool.

6. Distribute copies of the reproducible activity sheet on page 8. Follow the directions on the sheet.

Curriculum Crossovers:

1. The children can write stories about their shadows. Possible topics are:

   a. How their shadow became attached to them.

   b. The time their shadow got away or they lost it.

2. Help the children look through catalogs and find pictures of various window coverings. Look for some that are light-blocking such as dark shades, some that are light-filtering such as some vertical blinds. How much light does each material allow to pass through?

Take a Trip:

1. Encourage children to go on a moonlight walk with an adult. How many different shadows can they find? What is making each shadow? Where is the light coming from? Have the children share their moonlight walk information in class.

2. Visit the site of a sundial. Examine how the sundial works.

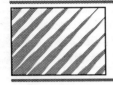

Shifting Shadows

Objects that block light cast shadows. Does a shadow always fall in the same place? Try this activity and see for yourself.

 What you need:
flashlight • paper cup • piece of white paper • a dark place

 What you do:

1. Look at the way the flashlight is being held in the picture. Where do you think the cup's shadow will fall? Circle the picture.

2. Now put the paper cup on the paper. Hold your flashlight in the same way. Where does the shadow fall? Circle the picture. Was your guess right?

3. Can you make the shadow fall in the places shown below? Use your flashlight. Mark an X over each picture to show where you held the flashlight to make each shadow.

Shadows on the Move: When you play outside, your shadow looks different at different times of the day. Why do you think this is so?

Nothing Sticks Like a Shadow

by Ann Tompert (Houghton Mifflin Co., 1984, F).

Rabbit loves playing with his shadow. Groundhog bets his brand new hat that Rabbit can't get rid of his shadow.

Activities:

1. Play "Whose Shadow Is It?" Place opaque objects on the stage of an overhead projector. Ask the children to name the objects that are making the shadows. Ask the children to bring in objects to place on the overhead for the shadow guessing game.

2. Have children go outside and observe one another's shadows as they run, jump, skip, etc. Then, with the children working in teams, have one shadow-maker in each team hold still while their partners outline his or her shadow with chalk or yarn. Compare the different shadows. Can someone else fit the outline of the team's shadow?

3. Have each child find a place to hide where no one can see his or her shadow. Ask the children to describe their special places. What is it about their special place that makes their shadow hard to find?

4. Have children use the reproducible worksheet on page 10 to make shadow puppets and put on a shadow show.

Curriculum Crossovers:

1. Tell a story about shadows or involving shadows.

2. Draw a rabbit or other animal playing with its shadow. "Act out" the story using a flashlight and hand shadows.

3. Collect newspaper and magazine photos that contain prominent shadows. Tape heavy paper over the object throwing the shadow. Can the children guess what the object is?

Take a Trip:

Talk a walk outside on a sunny day and feel different surfaces (in direct sunlight, in indirect sunlight, in light shade, in deep shade, etc.). How do they feel? Which surfaces are warmer? Children can use thermometers to find the temperature of each surface. Surfaces to try include grass, blacktop, cement, the ground under a tree, and a car.

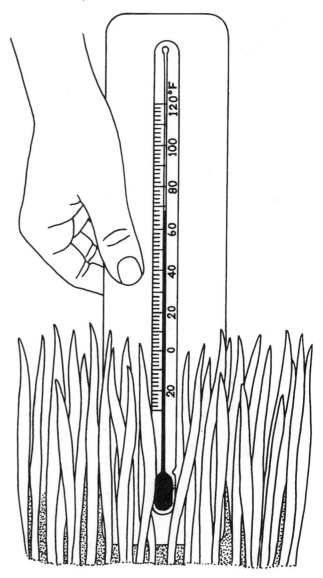

Shadow Play

Follow the directions to make these shadow puppets and put on a shadow show.

 **What you need:**
oak tag • pencil • scissors • a box • piece of white paper • tape • straws • lamp

To make the puppets:
1. Cut out each animal shape.
2. Trace the shapes onto the oak tag.
3. Cut out the cardboard animal shapes.
4. Tape a straw to the back of each animal.

To make the stage:
1. Cut a large "window" in the bottom of the box.
2. Tape the paper over the bottom of the box to cover the "window."
3. Turn the box on its side so that the "window" faces your audience.
4. Turn out the lights. Then turn on the lamp behind the box. Hold the puppets in front of the lamp. Where do the puppets' shadows fall?
5. Make up a story about the animals. Use the puppets to act out the story.

Challenge: How can you make the puppet shadows bigger? How can you make them smaller?

Sound

- *SHHHH!*, by Suzy Kline.
- *TY'S ONE-MAN BAND,* by Mildred Pitts Walter.

Background:

1. Sound is caused by vibrations in matter.
2. Vibrations can travel through objects. In particular, they can travel through air (and thus reach our ears).
3. The loudness of sound increases with the strength of the vibrations.
4. The pitch of sound depends on the vibrating material and its size.

Literature Tie-Ins:

- **SOUNDS** by David Bennett (A Bantam Little Rooster Book, 1989, NF)
- **NOISY NORA** by Rosemary Wells (Scholastic, 1973, F)
- **NOISY POEMS** by Jill Bennett (Oxford University Press, 1989, Poetry)

 ## Shhhh!

by Suzy Kline (Albert Whitman & Co., 1984, F).

A little girl is continually told by the adults in her world to SH!! Finally the SH-ing becomes too much for her. She goes outside and yells, screams, and makes all kinds of other noises before being quiet again.

Activities:

1. How are sounds produced? Have the children hold their hand on their larynx (throat) and say their name. Tell them that what they feel is a vibration. Ask the children to say the days of the week and months of the year, to growl, whisper, cough. Do all these sounds produce the same vibrations? Ask the children to describe the differences in the vibrations.

2. Give each student a rubber band. Ask the students to produce a sound using only the rubber band.

3. Have the children hold a rubber band tightly between the index fingers of both hands. They can rub their chin across the rubber band to produce a sound. What happens if they stretch the rubber band even more? What happens if they use a thicker rubber band? a thinner rubber band?

4. Secure a ruler or tongue depressor on the edge of a desk with the palm of one hand while gently plucking the free edge of the ruler with the other hand. Change the length of the protruding part of the ruler. Ask the children to describe how this changes the sound.

5. Strike a tuning fork with a rubber mallet or against your knee. (Never strike a tuning fork against a hard surface.) Place the prongs into a container of water. Ask the children to describe what they see.

6. Make an oscilloscope. Cut the ends out of a soup can and cover the rims with tape for safety. Stretch a balloon over one end. Hold the balloon in place with a rubber band. Stick a small mirror onto the balloon with white craft glue. Allow this to dry overnight. Hold the open end of the can to your mouth and cup your hands around it. Talk into the can while a partner shines a flashlight on the mirror. Your voice will cause the balloon to vibrate, and this vibration will be enlarged in the reflection of light on the wall.

7. Have the children look through magazines and cut out pictures showing objects that make sounds. Have them decide if they think

each object makes a loud or quiet sound. They can glue each picture in the appropriate column of the reproducible worksheet (page 13). Encourage the children to share their worksheets, giving reasons for where they placed each object.

Curriculum Crossovers:

1. In a creative writing story, use some of the sounds the little girl made.

2. Describe the quietest place you've ever been.

3. Brainstorm things that do *not* make sounds.

Take a Trip:

Go on a sound walk on the school grounds. No talking, only listening to the sounds all around. When you return, divide the class into groups and have each group list the sounds they heard. Combine the lists. How can the sounds be grouped? (For example, into natural and human-made, loud and soft, short and long, etc.)

Depending on where you live, a walk at different times of the year can give students the experience of different sounds, such as the rattle of dry leaves in the fall, the crunch of snow in winter, soft footsteps in spring, etc.

Loud or Quiet?

Cut out pictures from magazines that show things that can make sound. Decide if the sound is loud or quiet. Glue the pictures on the side of the paper where they best match.

Quiet	Loud

Ty's One-Man Band

by Mildred Pitts Walter (Scholastic, 1980, F).

Ty collects a washboard, a comb, two wooden spoons, and a pail for a stranger's one-man band.

Activities

1. Make (or help the children make) assorted musical instruments. Fill any cylindrical container with dried seeds, beans, etc. Make cymbals from pie pans, pot lids, etc.

2. Ask the children to place their fingers against their lips and gently blow and hum. Ask the children to describe what they feel and hear.

3. Demonstrate how sound travels through a solid. Tie two pieces of string to a spoon or metal coat hanger. Ask the children to wrap the strings around the tips of their index fingers. They should place the tips of their fingers near their ears. Next ask the children to gently tap the spoon against the table by rocking their bodies back and forth. What do they hear? Let the children try tapping other objects gently against the spoon. Different size spoons and spoons with different handle thicknesses will produce different pitches.

4. Set up an eight-note musical scale. Fill eight identical glass bottles with varying amounts of water (the first bottle 1/8 full, the next one 2/8 full, and so on.) Tap or blow across the opening of each bottle to produce a sound. Adjust water levels as needed. Add food coloring to the bottles so that students can easily distinguish and refer to the water level that produces each sound.

5. Use the reproducible worksheet on page 15 to have students make their own kazoos. Enjoy a kazoo concert!

Curriculum Crossovers:

1. Have the children sit quietly and listen to the sounds of their school and classroom. Ask the students to name some of these sounds. List the sounds they name.

2. Have the children brainstorm all the sound words they can think of.

3. Have students describe all of the sounds that they hear from the time they wake up till they arrive at school.

Take a Trip/Invite a Guest:

1. Visit a music shop and explore how various instruments make sounds.

2. Ask the instrumental music teacher (or other musician) to visit the class and play a variety of instruments for the students,

 # Strike Up the Band!

Would you like to make your own musical instrument? Just follow these directions to make a kazoo.

 ## What you need:
paper towel or toilet paper tube • wax paper • rubber band • pencil

 ## What you do:
1. Cover one end of the cardboard tube with a piece of wax paper.
2. Use the rubber band to hold the wax paper in place. Be sure the wax paper is stretched tight over the tube's opening.
3. Use the pencil to punch a hole in the tube just below the rubber band.
4. Hold the open end of the tube up to your mouth. Make sounds into the opening. Can you feel the wax paper vibrating, or moving?
5. Try to play a song with your kazoo by making sounds, such as "do-do-do," into the kazoo.

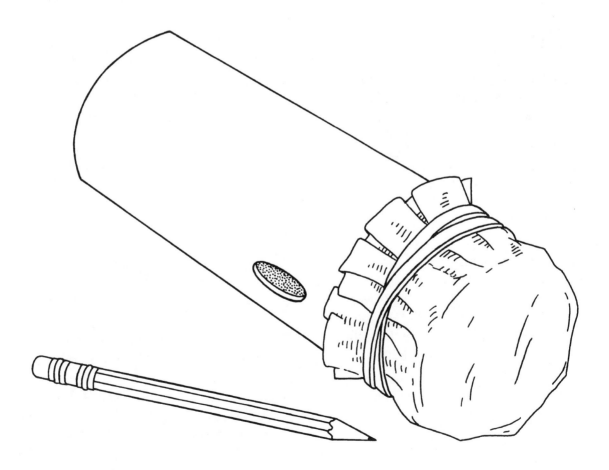

Simple Machines

- **THE CARROT SEED** by Ruth Krauss, illustrated by Crockett Johnson.

- **DR. DESOTO** by William Steig.

Background:

1. Levers, pulleys, inclined planes, and wedges are some examples of simple machines that make work easier.

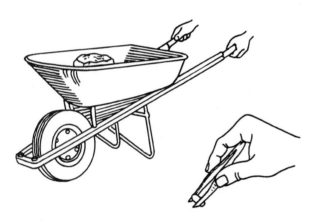

2. A lever is a bar that pivots on a fixed point called a fulcrum. It enables one to lift, move, or press different objects. Some levers (such as scissors and pliers) have the fulcrum between the object being moved and the force. Other levers (such as a wheelbarrow and a nutcracker) have the object between the fulcrum and the force. And still others (such as tweezers and a shovel) have the *force* between the object and the fulcrum.

3. A wheel turning on an axle or rod enables one to move objects smoothly from place to place.

4. A pulley is a wheel with a rope moving around it. By pulling on the rope one can conveniently raise objects.

5. An inclined plane is a slanting surface up which one can push heavy objects instead of lifting them.

6. A wedge is used to spread an object apart or to raise an object.

Literature Tie-Ins:

- **BATHTUBS, SLIDES, ROLLER COASTERS: SIMPLE MACHINES THAT ARE REALLY INCLINED PLANES** by Christopher Lampton (Millbrook Press, 1991, NF)

- **MARBLES, ROLLER SKATES, DOORKNOBS: SIMPLE MACHINES THAT ARE REALLY WHEELS** by Christopher Lampton (Millbrook Press, 1991, NF)

- **SEESAWS, NUTCRACKERS AND BROOMS: SIMPLE MACHINES THAT ARE REALLY LEVERS** by Christopher Lampton (Millbrook Press, 1991, NF)

- **KATY AND THE BIG SNOW** by Virginia Lee Burton (Houghton Mifflin Co., 1943, 1971, F)

- **MIKE MULLIGAN AND HIS STEAM SHOVEL** by Virginia Lee Burton (Houghton Mifflin Co., 1939, 1967, 1976, F)

 ## The Carrot Seed

by Ruth Krauss, illustrated by Crockett Johnson (Harper & Row, 1945, F.).

A little boy plants a carrot seed and everyone tells him "it won't come up." But the little boy weeds and waters it regularly. One day it comes up just as he thought it would. It is so big that it requires a machine (wheelbarrow) to carry it away.

Activities:

1. If possible, bring a wheelbarrow to class. Select a heavy object and have students try moving it about. Then place the object in the wheelbarrow and have them move it about. Have them compare how it felt to move the object with and without the wheelbarrow.

2. Use a thick board (lever) and a brick (fulcrum) to enable a student to lift you. An ideal board is hardwood such as a leaf from an old table, but a two-by-four will work. Place the brick under the board near one end. You then stand on that end of the board. Have the student place a foot on the other end of the board and press down. Use other students as spotters, standing close to you and the student for safety. You and the student can place your hands on the spotters' shoulders for stability.

   After the student has lifted you, move the brick closer to the center of the board. Have the student lift you again. Have the student describe how it felt to lift you each time.

   The brick can be moved to other positions and students can try to lift you again. The closer the brick is to the load being moved, the easier it is to move.

3. Have the students try cracking a nut with their hands. Ask them to describe how it feels. Provide them with nutcrackers. Have them predict where the nut should be placed for the least amount of force to be used in opening it. Have them test their hypotheses. Have them compare the effectiveness of the nutcracker and their bare hands. Point out that, unlike the board in Activity 2, the nutcracker has its fulcrum at the end.

4. Make a force measurer (see the diagram below). Attach a rubber band to one end of a ruler with reinforced tape. Open a paper clip and hook one end through the rubber band. Use the force measurer to pull an object (such as an eraser or a film can filled with dirt or sand) across the desk. Measure the amount of force needed by determining how far the rubber band is stretched along the ruler. Use the reproducible data sheet on page 18 to record the amount of force used. Place the object on a toy car with wheels and axles. Measure the force needed to move the object now. Ask the children to compare the forces required with and without wheels and axles.

Curriculum Crossovers:

1. Brainstorm types of levers that might be found at home. Make a list. Ask parents to work with children to find those on the list as well as others. Have children compare their lists.

2. Have students describe to the class (without showing) a lever they found at home. Ask the other children to guess what it is used for.

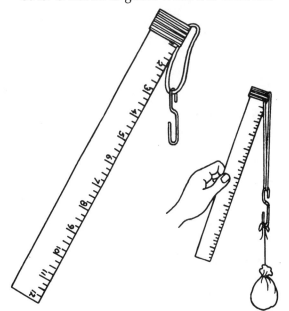

Take a Trip/Invite a Guest:

1. Visit a museum that has a collection of old tools. Request a docent to explain how they were used and what has replaced them.

2. Ask a carpenter to visit the class with his or her tools. Have the carpenter demonstrate how much easier work is when one has the right tool and knows how to use it correctly.

How Much Force?

Use the force measurer you made in class to find out how much force it takes to move different objects. Use the chart below to record the different amounts of force.

Object	Amount of force needed to move:	
	with wheels	without wheels

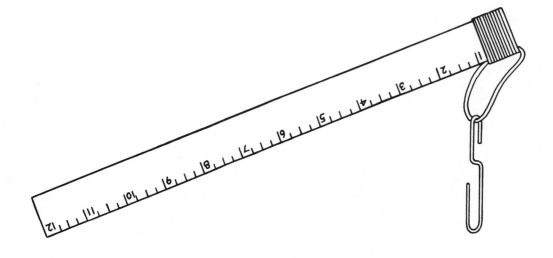

Dr. DeSoto

by William Steig (Farrar, Straus and Giroux, 1982, F.).

Dr. DeSoto is a dentist of excellent reputation. Animals of all sizes come to him with their toothaches. He reaches into the mouths of large animals with a series of pulleys. Dr. DeSoto almost has a problem with a fox but outfoxes him.

Activities:

1. Construct pulleys from empty thread spools (see the diagram on page 20).

Test the fixed pulley:

   a. Attach the fixed pulley to a meter stick or an equivalent support.

   b. Thread it with twine.

   c. Attach the container of sand to one end of the twine.

   d. Attach the force measurer to the other end of the twine and pull.

   e. How far does the rubber band stretch? Record the amount of force needed to raise the sand.

Compare the movable pulley with the fixed pulley:

   a. Attach one end of the twine to the meter stick.

   b. Thread the twine through the pulley.

   c. Attach the container of sand to the pulley.

   d. Attach the force measurer to the other end of the twine and pull.

   e. How much force was needed to raise the sand? Record the amount of force needed. Then compare the amount of force needed for the fixed and the movable pulleys.

2. Try to recreate the secret formula that Dr. DeSoto used to seal Mr. Fox's mouth. A recipe for paste: In a saucepan, slowly mix 1 cup of water into 1/2 cup of rice flour or wheat flour. Bring to a boil over low heat. Stir until thick and glossy, about 5 minutes. What can you paste shut?

Curriculum Crossovers:

1. Brainstorm ways in which fixed or movable pulleys might be used. Encourage the children to come up with amusing or fantastic ideas.

2. How many dentists are there in your town? Where would you find that information?

Take a Trip/Invite a Guest:

1. Take the class on a pulley scavenger hunt. Have them look carefully for pulleys. When you return to class ask: Where did you see pulleys? How were they used? Are there other things that could have been used? How do they make life easier?

2. Take a trip to a hardware store and investigate the types of pulleys sold. Read the packaging to find out how they are used.

3. Invite a veterinarian to visit the class and tell of his experiences with animals that have tooth problems.

Pulley Pull

You can measure the force needed to lift various objects by making a force measurer with a rubber band, a ruler, a piece of string, an empty thread spool, a piece of wire, and an opened paper clip. Use the diagram below, and ask your teacher for help.

Directions: Attach the object you want to lift to the end of the string. (Point A in the diagrams below.) Attach the other end of the string to opened paper clip. (Point B in the diagrams below.)

⬤ **If using the fixed pulley, pull down on the force measurer.**

⬤ **If using the movable pulley, pull up on the force measurer.**

Always be careful to keep the rubber band and paper clip against the ruler. Once the object is lifted, use the chart below to record the point where the paper clip is. (The number on the ruler that's closest to the paper clip.) Write the number in the correct column below. Try moving a lot of different things. Which take the most force to move? The least?

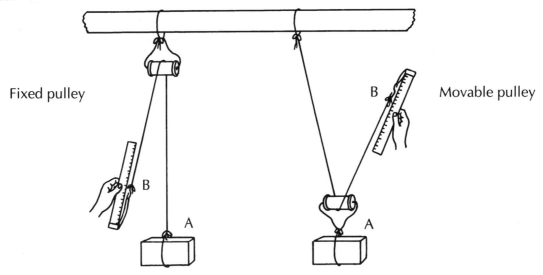

Object:	Force needed:	
	For a movable pulley	For a fixed pulley

Time

- **THE GROUCHY LADYBUG** by Eric Carle.
- **CLOCKS AND MORE CLOCKS** by Pat Hutchins.

Background:

1. In our culture, schedules require people to be able to tell time.

2. Yesterday, today, and tomorrow are the first concepts of time children learn.

3. Later, children learn more specific divisions of time: seconds, minutes, hours, days, weeks, months, years, etc.

4. The passage of time during the day can be measured approximately by the apparent movement of the sun across the sky.

5. The simplest mechanical device for measuring time is the pendulum, the basis of most early clocks. The period of a pendulum (the time it takes to complete a swing) depends on its length, not on the distance of the swing. Thus a pendulum can be made to measure a fixed period of time, such as one second.

Literature Tie-Ins:

- **CLOCKS AND HOW THEY GO** by Gail Gibbons (Thomas Y. Crowell, 1979, NF)

- **KNOW ABOUT TIME** by Henry Pluckrose, photography by Chris Fairclough (Franklin Watts, 1988, F)

- **NOBODY HAS TIME FOR ME** by Vladimir Skutina (Wellington Publishing, Inc., 1988, 1991, F)

- **THE SCARECROW CLOCK** by George Mendoza, illustrated by Eric Carle (Holt, Rinehart and Winston, F)

- **TIME TO...** by Bruce McMillan (Lothrop, Lee & Shepard Books, 1989, NF)

The Grouchy Ladybug

by Eric Carle (Harper & Row, Publishers, 1977, F).

At five o'clock in the morning, Grouchy Ladybug arrives at a leaf covered with aphids (very small insects that suck juices from leaves) at the same time as a Friendly Ladybug. Grouchy tells her to go away or fight for the food. But when Friendly Ladybug challenges her, Grouchy declares that she is too small to fight and flies off. Each hour throughout the day she encounters a new and larger animal but each time she decides not to fight because, she says, the animal is too small. At six in the evening, a very tired and hungry Grouchy Ladybug returns to the aphid-covered leaf.

Activities

1. To help the children develop a feel for a minute, have them sit quietly while you time a minute. Afterward, ask the class if they were surprised at how long a minute was. Ask if they think they can now tell how long a minute is. With the class unable to see a clock, ask them to snap their fingers, stopping only when they think a minute has passed.

2. Have the children suggest pairs of similar things, one of which moves slowly and the other quickly. For example: fish in an aquarium and a snail; the students getting ready for bed and getting ready to go out to play; or honey and water dripping from spoons. Then have them measure and compare the length of time it takes to go from the classroom to a certain class; get ready for lunch; tie a shoelace; go home from school, and so on.

3. Have each student keep a schedule of starting and stopping times of common activities they perform at home, such as eating, homework, chores, watching TV, practicing piano (or other instrument), bathing, getting ready for bed, going to bed.

4. Distribute copies of the reproducible data sheet on page 23, and complete as a class activity, or assign for homework.

Curriculum Crossovers:

1. On strips of paper, measure off the size of some of the animals that Grouchy Ladybug encounters. Have the children draw a picture of the animal on the appropriate length strip, then find items in the class that are near the size of each animal.

Firefly	**15 mm**
Aphid	**8 mm**
Ladybug	**8 mm**
Yellow jacket	**25 mm**
Stay Beetle	**4 cm**
Praying Mantis	**10 cm**
House Sparrow	**15 cm**
Lobster	**50 cm**
Skunk	**45 cm**

2. Write a class story using the different words that describe Grouchy Ladybug meeting the different animals: flew, met, saw, came across, bumped, spotted, happened upon, ran into, and encountered.

3. Encourage the children to move like the animals in the book. Ask the children to describe how they feel when they move like different animals.

Invite a Guest:

1. Invite a gardener or naturalist to the classroom to discuss the natural clocks in plants and animals. For example, marigolds flower at 7 a.m., evening primroses around 6 p.m..

2. Invite a gardener or naturalist to the classroom to discuss the benefits of ladybug beetles in the garden.

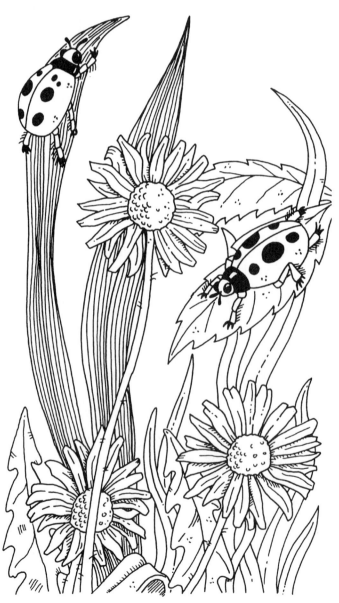

All in a Day

How does your day compare with the grouchy ladybug's day? Draw the hands on the clock to show each time. Then fill in the chart to show your daily schedule.

Time		Grouchy's Day	Your Day
8 a.m.		Grouchy meets a praying mantis	
10 a.m.		Grouchy meets a lobster	
12 p.m.		Grouchy meets a boa	
2 p.m.		Grouchy meets a gorilla	
4 p.m.		Grouchy meets an elephant	
6 p.m.		Grouchy meets the friendly ladybug for the second time	

Clocks and More Clocks

by Pat Hutchins (Macmillan, 1970, F).

Mr. Higgins finds an old grandfather clock in the attic. To see if it tells correct time, he purchases another clock and puts it in the bedroom. But as it takes him time to go from room to room, he finds that the two clocks don't agree. So he buys another and then another, but still has the same problem. Mr. Higgins finally decides to ask a clock master to look at his clocks. The clock master checks each with his pocket watch and finds that they all keep perfect time. After Mr. Higgins purchases a pocket watch, he too finds them to be in agreement.

Activities:

1. Display clocks and pictures of clocks. Encourage students to contribute to the collection. Ask: How are they alike? How are they different? Do any of them have a characteristic that is alike? Help the children group them by different characteristics, such as power sources (electric current, battery, or mainspring), digital or analog, types of numerals (if any): Roman or Arabic, chiming or not, with or without a second hand, with or without a pendulum, etc.

2. From a paper plate and two strips of card make a picture clock showing activities the children do during the day. Select children to move the clock hands when starting a new activity.

3. Construct a shadow clock. Set a pencil in a lump of clay and put it in the center of a large piece of paper placed in full sunlight. At even intervals throughout the day draw the shadow created by the pencil and record the time beside the shadow.

4. Set up a pendulum as shown on the reproducible activity sheet on page 25.

   a. Start the pendulum swinging and observe. Count how many pendulum periods it makes in 15 seconds.

   b. Add more weight to the bob (the weight at the end of the pendulum) and count again.

   c. Use a longer piece of string and count again.

   d. Use a shorter piece of string and count again.

Curriculum Crossovers:

1. Ask children to look for different kinds of clocks in their homes and neighborhoods (in stores, on public buildings, etc.). Then have them write about or describe orally the most interesting clock they saw.

2. Have students write about a day without clocks.

3. Have students write a story from their favorite clock's point of view.

Take a Trip/Invite a Guest:

1. Go to the playground and use the swings to measure the number of periods (complete swings to and fro) in 30 seconds. Compare the number of periods between different children. If possible, adjust the length of the swings. A swing is like a pendulum in that it goes back and forth and the length of its chain determines the period.

2. Invite a clock repair person to the class to show the inside of a clock. Have him or her demonstrate how the mechanism functions.

Name

Pendulum Power

You can make your own pendulum with a pencil, some tape, a piece of string, an opened paper clip, and something to use as a weight.

Directions: Start the pendulum swinging and count how many periods (complete swings back and fourth) it makes in 15 seconds. Write the type of weight and the length of string in the correct column below, and the number of swings in the column next to it. Experiment by changing the weight and the length of the string.

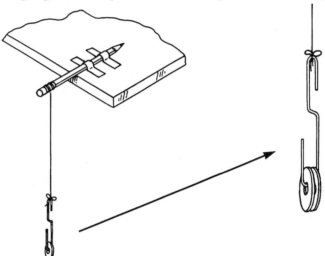

Weight	Swings in 15 Seconds

Length of String	Swings in 15 Seconds

Seeds and Plants

- *THE REASON FOR A FLOWER* by Ruth Heller.

- *THE BIGGEST PUMPKIN EVER* by Steven Kroll.

Background:

1. Seeds house dormant "baby" plants.

2. Different plants have different seeds that can grow only into their own kind of plant.

3. A bulb is an underground stem.

4. The fruit of a plant is where seeds for new plants are produced.

5. If all seeds fell straight to the ground, they would compete with one another and with the parent plant for soil nutrients and sunlight. Seeds travel in many ways.

6. Seeds that are planted in soil, are kept moist, and receive sunlight can become mature plants.

Literature Tie-Ins:

- **THE TINY SEED** by Eric Carle (Scholastic, 1989, F)

- **PUMPKIN-PUMPKIN** by Jeanne Titherington (Scholastic, 1986, F)

- **KIMI AND THE WATERMELON** by Miriam Smith (Puffin books, 1983, F)

- **THE CARROT SEED** by Ruth Krauss (Harper & Row, 1945, Scholastic Big Book, 1989, F)

- **A SEED IS A PROMISE** by Claire Merrill. (Scholastic, 1973, F)

The Reason for a Flower

by Ruth Heller (Grosset & Dunlap, 1983, NF).

Describes the stages in the development of all kinds of flowers from pollination to fruiting. Also explains how seeds travel and grow in different places, and how flowering plants provide people and animals with food and many other benefits.

Activities:

1. Have the children collect a variety of seeds. (Enlist the help of parents.) Who can find the biggest seed? the smallest seed?

2. Have the children use magnifiers to observe the seeds. Children can observe, compare, and describe colors, sizes, shapes, how each seed feels, what happens when the children gently blow on each seed, and what happens when they try to roll each seed.

3. Make a "See How We Grow" bulletin board. Fold a piece of paper towel in fourths. Wet the paper towel and squeeze out the excess water. Place the square in a sealable plastic bag. Place assorted seeds between the paper towel and the bag. Partially seal the bags. Use a thumbtack to attach them to a sunny bulletin board.

4. Give each child an unshelled peanut. Ask the children to open the shell and find the nut (seed). Ask them to carefully open the nut and find the tiny plant between the two halves of the nut. How many other examples of "Seeds We Eat" can the children find?

5. Have children complete the reproducible worksheet on page 28. They can then cut it into sections and make it into a flip book.

Curriculum Crossovers:

1. Purchase hardy bulbs in the fall. Crocuses, hyacinths, tulips, daffodils, and narcissus can be forced in school. By the end of October, plant two to three bulbs each in 6-inch pots. Bury the bulbs to the tip in perlite or vermiculite, or a mixture of 1/3 commercial potting soil, 1/3 vermiculite, and 1/3 peat moss. (These ingredients are obtainable from a nursery or garden supply store.) Water the bulbs, cover each pot with a plastic bag, and set it in a cold place. The cold will simulate winter temperatures, which allow for root development. After eight weeks move the containers into the classroom. The bulbs should bloom in 2-4 weeks.

After you and the children have enjoyed the bulbs in bloom, cut off the flower heads and allow the leaves to brown naturally. Add fertilizer regularly. After the leaves have turned brown, plant the bulbs outside. Note: The bulbs may not bloom again for two or three years.

2. Grow an edible sprout garden. Place a piece of damp paper towel on a dish or other flat surface and sprinkle it with seeds such as wheat, mustard, cress, mung, alfalfa, and radish. After the seeds sprout, allow the children to taste the sprouts. Do they all taste the same? Which do the children like best?

3. After discussing some of the different ways seeds travel, children can pretend to be seeds and describe their journey to the place where they choose to land and grow.

Take a Trip

Have the children cover one shoe with an old sock. Take a walk in an unmowed, overgrown field. Have children look for hitchhikers (seeds) on their socks. How many seeds did they pick up? Lightly cover the sock with soil and see what grows.

Little Sprout

Directions: Color each picture. Cut on the dotted lines. Put pages in number order. Staple and flip. Fill in the blanks under each picture and tell something about your plant.

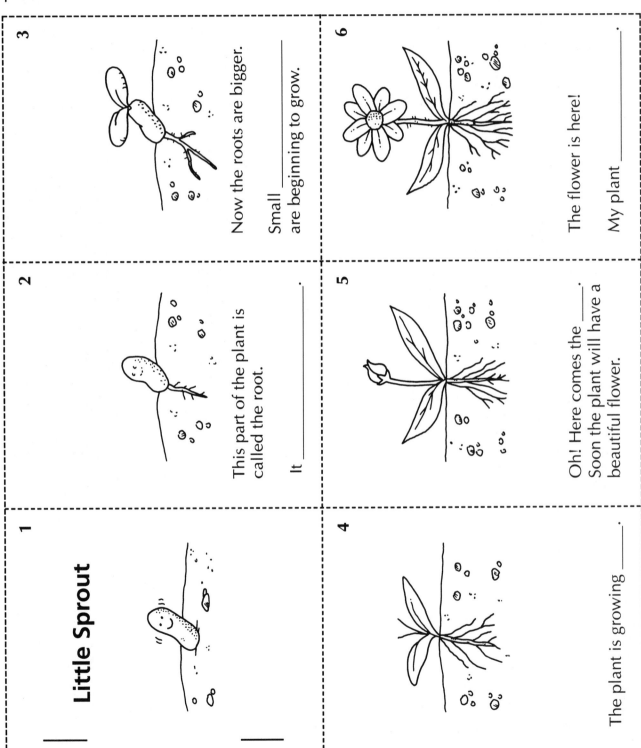

3

Now the roots are bigger.

Small _____ are beginning to grow.

6

The flower is here!

My plant _____ .

2

This part of the plant is called the root.

It _____ .

5

Oh! Here comes the _____ . Soon the plant will have a beautiful flower.

1

Little Sprout

_____

4

The plant is growing _____ .

The Biggest Pumpkin Ever

by Steven Kroll. (Scholastic, 1984, F).

Two mice fall in love with the same pumpkin. Clayton wants his to grow big enough to win the grand prize in the pumpkin contest. Desmond wants his to make the biggest jack-o'-lantern in the neighborhood.

Activities:

The pumpkin is a wonderful subject for observation, prediction, measurement, and problem solving.

1. Children can first predict and then measure or count:

   a. how heavy the pumpkin is;

   b. how tall;

   c. how big around;

   d. how many ribs it has;

   e. the thickness of its skin;

   f. the thickness of its meat;

   g. the number of scoops of seeds it contains. (Use coffee scoops or any similar measures, making sure they are the same size for each student.)

2. Children can describe:

   a. the color of the inside and outside of the pumpkin;

   b. the appearance of the inside and outside;

   c. the size, color, and shape of the seeds.

3. Children can observe the pumpkin and determine:

   a. if the ribs on the pumpkin correspond to the ridges on the stem;

   b. the purpose of the bottom of the pumpkin.

4. Ask the children to predict what will happen when a pumpkin is placed in a container of water. Do they think it will float or sink? What position will it be in: stem up? stem down? stem to the left? to the right? Have the children draw their predictions on the reproducible worksheet (see page 30). Then actually place a pumpkin in a water tank and see the results. Ask the children to guess if all pumpkins float the same way. Try other fruits or vegetables. Will they float or sink? What position will they be in?

Curriculum Crossovers:

1. Children can draw a picture of what they think the inside of a pumpkin will look like when it is cut open.

2. If there is a toaster oven at your school, you can roast the pumpkin seeds for children to eat.

3. Children can grow their own class pumpkin plants. Rinse off a scoop of pumpkin seeds and allow them to dry. Then place them either in a paper cup with a little soil or on a piece of damp paper towel in a sealable plastic bag.

Take a Trip:

1. Visit a pumpkin patch. Adopt a pumpkin plant and chart its growth.

2. Visit a bakery and learn about making pumpkin pies.

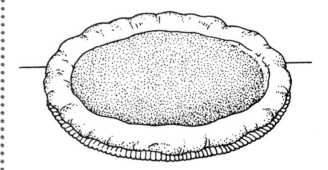

Name

Float or Sink

What do you think will happen when you place a pumpkin in a tank of water? Will it float or sink? Stem up or stem down? Draw what you think will happen in the tank. Compare it with what **really** happens when you place a pumpkin in the tank in class.

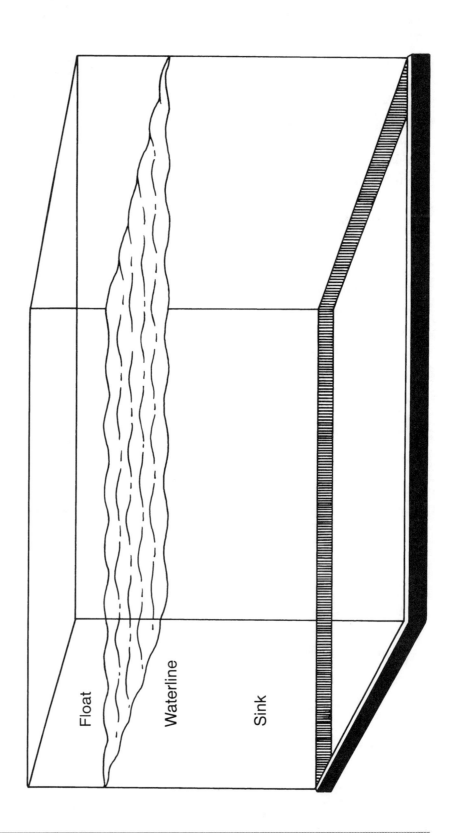

Float

Waterline

Sink

Trees

- *MIGHTY TREE* by Dick Gackenbach.

- *RED LEAF, YELLOW LEAF* by Lois Ehlert.

Background:

1. Trees differ from other plants in having a tough woody stem.

2. Trees have roots, trunks, and branches. Many trees have leaves, flowers, and fruits or nuts.

3. The bark of a tree protects it from drying out and from attack by insects.

4. Some trees lose their leaves in the fall (they are deciduous).

5. Some trees are green throughout the year (they are evergreens).

6. Different trees have leaves of different sizes, shapes, and textures.

7. Trees provide people with many products, including fruits, nuts, wood, and paper.

8. Trees provide clean air, shade, and beauty.

Literature Tie-Ins:

- **THE GIFT OF THE TREE** by Alvin Tressett (Lothrop, Lee Shepard Books, 1992, NF)

- **ONCE THERE WAS A TREE** by Natalia Romanova (Dial Books for Young Readers, 1985, F)

- **A LOOK INSIDE A TREE.** by Gina Ingogli (Grosset & Dunlap, 1989, NF)

- **A TREE IN A FOREST** by Jan Thornhill (Simon and Schuster Books for Young Readers, 1992, NF)

Mighty Tree

by Dick Gackenbach (Harcourt Brace Jovanovich, 1992, F).

Three seeds become three trees. The "most important tree of all" still stands in the forest today.

Activities

1. Adopt a tree to study throughout the year. Have the children keep a journal. It could be a photo journal or a picture journal. The children can describe the tree and its surroundings. Have them observe the tree carefully. Ask questions to guide their discovery. Can they reach around the tree? What does the bark feel like? Are all the branches on the tree the same? How would they describe the tree's leaves? How do the leaves of their tree change throughout the school year? Are there any signs of animal activity around the tree?

2. As the trunk of a tree grows, the bark also grows and stretches like skin. Children can make a bark rubbing by placing a large piece of paper against the bark of a tree and rubbing over the sheet of paper with a crayon. Do the same with leaves. Have the children compare their rubbings. Is the bark of all trees the same? How many different kinds of bark can they find?

3. Have the children cut and paste a piece of their bark and leaf rubbings to the outline of their tree (see the reproducible worksheet on page 33). They can also add drawings of other special features of their tree, including anything that lives in, on, or near the tree. They can do these activities seasonally, to record the changes in their tree throughout the year.

4. Help children find out how old a tree is by counting the rings in a slice of its trunk. (Use 5 cm thick slices from the trunk of a small fallen or discarded tree such as a Christmas tree.) It is easiest to count the dark rings which show the end of a year's growth. Children can rub the slices across sandpaper to smooth the surface and make the rings easier to see. A thin coat of paste wax will also make the rings more visible.

5. Demonstrate how water moves up through the plant to the leaves by placing one leafy stalk of celery in a tall glass of plain water and another leafy stalk in a glass of colored water. Have the children observe the changes in the celery leaves after a day. How can they explain the changes?

6. Children can better understand the help we get from shade trees if they go outside on a warm, sunny day and take the air temperature in a sunny spot and under a shade tree. Ask them to explain the difference in the temperatures.

Curriculum Crossovers:

1. Ask each child to bring in a leaf from any tree of their choice (preferably a fallen leaf). Have the children compare their leaves. Are all the leaves alike? How do they differ?

2. Children can pretend that they are trees and describe the most unusual "visitor" they ever had.

3. Have the children brainstorm — or find in the classroom — as many objects as they can that come from trees. You might prompt them by asking, for example, what paper is made from.

Take a Trip/Invite a Guest:

1. Take a walk in an area with trees. How many different kinds of trees can you find? How many tree sprouts can you find?

2. Visit a Christmas tree farm or plant nursery. Are all trees alike? Compare their needles, leaves, and bark.

3. Contact your local government to find the location of the nearest Shade Tree Commission. Invite a representative of the Commission to visit your class to discuss how and why its members choose the types of trees that they are going to plant.

Our Tree!

Directions. Glue a piece of your bark rubbing to your tree. Glue a piece of your leaf rubbing to your tree. Draw everything you see on or around your tree!

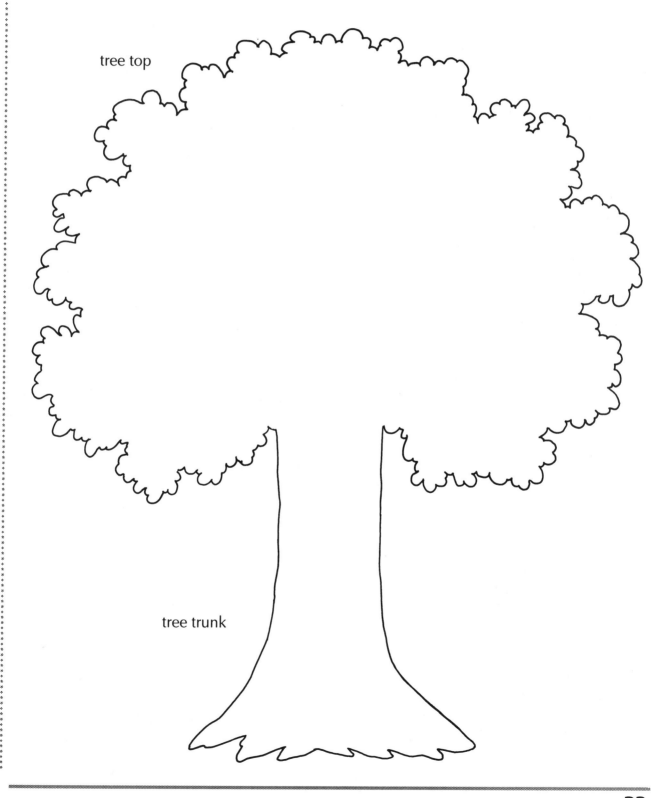

tree top

tree trunk

Red Leaf, Yellow Leaf

by Lois Ehlert (Harcourt Brace Jovanovich, 1991, F).

A young child tells the story of the growth of his sugar maple tree.

Activities:

1. Start a class leaf collection. Include leaves collected from your adopted tree (see Activity #1 under *Mighty Tree*, page 32) during each of the seasons. You can dry leaves between pieces of paper towel or newspaper and preserve them between sheets of clear contact paper. They keep their color for years.

2. Compare the size and shape of several leaves from different locations on your adopted tree tree by tracing the outlines of the leaves on graph paper. Counting squares on the graph paper is an easy way to compare size. Children can count any squares that are half or more covered.

3. Compare the leaves from your adopted tree to leaves from different trees. Can the children find leaves that are egg shaped, heart shaped, long and pointed, short and pointed, narrow, wedge shaped? Can they find leaves with smooth edges and leaves with toothed edges? How many different color leaves can they find?

4. Make leaf rubbings. Place leaves bottom side up. Lay a piece of paper over the leaves and rub gently with a crayon.

5. Use the reproducible worksheet on page 35 to have students write stories about the cycle of a tree's yearly growth.

Curriculum Crossovers:

1. Pretend you are a tree and describe one day in your life. An ongoing assignment might be to describe your life throughout the school year.

2. The children can collect and sprout acorns in the fall. To sprout the acorns they can use either sealable plastic bags or soil. With the former, they simply place one or two acorns in each bag and seal it. With the latter, they place one acorn in a plastic cup with a little soil. They then store the bags or pots in a dark location. Have the children observe their bags or pots weekly and note any changes.

3. Determine the approximate height of your tree. Stand in front of the tree and hold a pencil out, upright, at eye level. Walk away from the tree slowly until the top and bottom of your pencil line up with the top and bottom of the tree. Keeping the pencil the same distance from your eyes, turn it on its side so that the eraser lines up with the base of the tree. Ask a partner to stand on the ground at the point that lines up with the tip of the pencil. Pace off the distance between the tree and your partner. (Or measure the distance with a meter stick or measuring line.) This distance is the height of your tree. How does it compare to the height of other nearby trees?

Take a Trip:

If you live in an area where the leaves turn color in the fall, take a walk in the woods during that season. Walk in the same area again in the spring when the trees are in bloom and leafing out. Ask the children to describe what they see.

Name

The Seasons of a Tree

Look at the trees on this page. Decide which season each tree is in by looking at the branches. Write a sentence to tell what happpens to each tree in each season.

_____

_____

35

Apples

- *WHO WAS JOHN CHAPMAN?* by Patsy Becvar.

- *THE SEASONS OF ARNOLD'S APPLE TREE* by Gail Gibbons.

Background:

1. The apple tree is one of the most popular fruit trees in America. Apples are used in more products than any other fruit.

2. There are thousands of kinds of apples, of which a few dozen are used commercially. Each of these has a slightly different color, shape, or taste.

3. The apple is classed as a pome — a firm-fleshed fruit with a core containing several seeds. It thus differs from fruits such as cherries and peaches (which contain a single seed), strawberries and raspberries (which have many seeds), and so on.

Literature Tie-Ins:

- **THE LIFE AND TIMES OF THE APPLE** by Charles Micucci (Orchard Books, 1992, NF)

- **AN APPLE TREE THROUGHOUT THE YEAR** by Claudia Schnieper (Carolrhoda Books, Inc., 1984, F)

- **APPLEMOUSE,** translated by Timothy Cleary (Hill and Wang, 1984, F)

- **JOHNNY APPLESEED.** by Eva Moore (Scholastic, 1964, F).

- **APPLE TREE** by Peter Parnell (Macmillan, 1987, F)

- **APPLE TREE CHRISTMAS** by Trinka Noble (Dial Books, 1984, F)

Who Was John Chapman?

by Patsy Becvar (Nystrom, 1991, F).

A class tries to determine whether or not Johnny Appleseed was a real person.

Activities:

1. Using a variety of apples, compare and contrast their outsides and their insides. Things to observe include: color, shape, size, mass, volume, taste, and number of seeds (see the reproducible worksheet, page 38).

2. Measure the height and circumference of apples. Find their weight. Who can find the biggest apple? the smallest apple?

3. Collect data on the number of seeds found in different species of apples. Draw the size and shape of the seeds.

4. Have a taste test to see which apples are favored when raw. Remove the skin and slice the apples first, so that the pieces look alike.

5. What other kinds of fruit are most like apples? What makes them alike? What kinds are very different from apples? What makes them different?

Curriculum Crossovers:

1. Make an apple recipe book. Have the children bring in their family's favorite apple recipe.

2. Make apple butter or apple sauce. Here's an apple butter recipe:

   Quarter and remove stems and seeds from 4 pounds of Jonathan or Winesap apples. Cook in 2 cups of cider or apple juice until soft. Press through a fine strainer.

For each cup of pulp add 1/2 cup of brown sugar. To the mixture add 1 teaspoon cinnamon, 1/2 teaspoon cloves, and 1/4 teaspoon ground allspice.

Cook over low heat, stirring constantly until the sugar is dissolved. Continue to cook for at least 2 hours, stirring frequently. Test by putting a spoonful on a plate. When no rim of liquid separates around the mixture, it is done.

3. Write a letter to the International Apple Institute at 2430 Pennsylvania Ave., NW, Washington D.C. 20037. Ask how many varieties of apples are grown in the U.S. and which states produce the greatest quantity of apples.

Take a Trip:

1. Visit a supermarket and identify the varieties of apples available. Ask the produce manager to explain where the apples come from.

2. Visit a cider press and observe how cider is made.

Name

Inside an Apple

Directions: Draw and color a picture of each type of apple you look at. How many seeds did you find inside?

1. Number of Seeds_____	**2.** Number of Seeds_____
3. Number of Seeds_____	**4.** Number of Seeds_____

To Think About: If you were a farmer which of these apple trees would you like to grow? Why? _____

_____

The Seasons of Arnold's Apple Tree

by Gail Gibbons (Harcourt Brace Jovanovich, 1984, NF).

A book about the activities of an apple tree and a little boy during each of the four seasons.

Activities:

1. Sprout apple seeds. All apple seeds except red and yellow Delicious will work. Dry apple seeds for 1-2 weeks. Place a dampened paper towel into a sealable plastic bag. Place the seeds in the bag so that they rest on the paper towel. Place the bag in the freezer for two months. Then take the seeds out of the freezer and plant them in approximately one inch of potting soil. Use a planter that has good drainage. Place in a warm and sunny spot and water regularly.

2. Find the "magic star" inside an apple. Place the apple sideways on the cutting surface and cut through the middle. (For safety, only an adult should do the cutting.) Separate the two halves. On each half you will see a five pointed star.

3. You can preserve your "magic stars" by making thin cuttings from the center of each apple and allowing these to dry in the classroom. Do any other fruits have magic stars? Do all varieties of apples have stars? Do the stars always look the same?

4. Have children color the illustrations on the reproducible worksheet (page 40) and then place them in sequence.

Curriculum Crossovers:

1. Write to a doctor and ask if an apple a day really keeps the doctor away.

2. Draw a picture of a tree that could grow from an apple seed.

3. Cut open or slice various apples, coat the inner surface with tempera paint, and make apple prints.

Take a Trip:

Visit an apple tree in blossom. Observe the apple tree throughout the year. How does it change?

Magic Star

1.

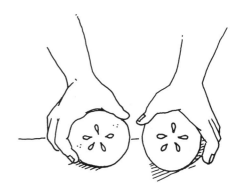

2.

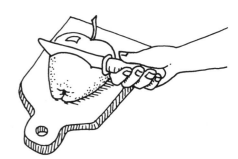

3.

From Flower to Fruit

Directions: Color each picture. Cut on the dotted lines. Put them in order from flower to fruit.

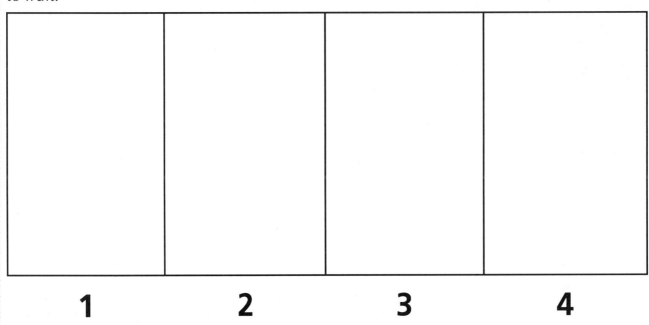

1　　　　**2**　　　　**3**　　　　**4**

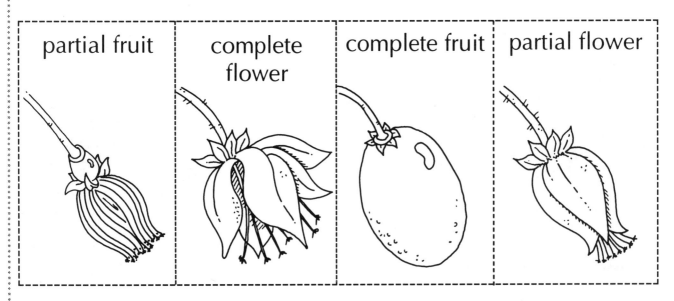

partial fruit　　complete flower　　complete fruit　　partial flower

Bite In: Pretend you have just picked the apple off the tree's branch. Bite into it. Use all of your senses to tell about what eating the apple is like. How does it sound, taste, smell, and so on? You can write on the back of this paper.

or

Fanciful Fruit Tree: You know that different fruits grow on different trees, plants, and bushes. Just for fun, pretend that all fruits come from one tree. What would the tree look like? Draw a picture of it on the back of the page. Add as many fruits as you can think of.

Seasons

- *CHIPMUNK'S SONG* by Joanne Ryder.

- *A BUSY YEAR* by Leo Lionni.

Background:

1. In our spring and summer, Earth's Northern Hemisphere is tilted toward the sun. We get more direct light and heat from the sun during these seasons.

2. In our fall and winter, the Northern Hemisphere is tilted away from the sun. We get less direct light and heat during these seasons.

3. Animals and plants respond differently to the changes in the seasons. Some animals get ready for winter by migrating (moving to a new home). Others hibernate (sleep for the winter).

4. Some trees lose their leaves in the winter, blossom or leaf out in the spring, and produce fruit or nuts in the fall.

Literature Tie-Ins:

- **SUNSHINE MAKES THE SEASONS** by Franklyn M. Branley (Thomas Y. Crowell, 1985, NF)

- **MOUSEKIN'S WOODLAND SLEEPERS** by Edna Miller (Prentice-Hall Books, 1970, F)

- **SEASONS by David Bennett** (A Bantam Little Rooster Book, 1988, NF)

- **SEASONS by Melvin Berger** (Doubleday, 1990, NF)

- **WINTERTIME** by Ann Schweninger (Viking, 1990, NF)

- **WINTER** by Ron Hirschi (Cobblehill Books, NF)

- **THE YEAR AT MAPLE HILL FARM** by Alice and Martin Provensen (Atheneum, 1978, NF)

- **A BOOK OF SEASONS** by Alice and Martin Provensen (Random House Picturebook, 1976, NF)

- **UNDER YOUR FEET** by Joanne Ryder (Macmillan, 1990, F)

 ## Chipmunk's Song

by Joanne Ryder (Lodestar Books, Dutton, 1987, F).

A small boy shrinks so that he can experience the life of a chipmunk through the four seasons. A close-up look at the home and habitat of a chipmunk.

Activities:

1. Conceal the cover of the book while reading the story to the class. Ask them to guess what animal the story is about. Ask them to draw a picture of that animal. Have them hold up their pictures when you have finished reading the story.

2. Have children who own pets with fur (pets who spend time outdoors are better) brush the pet 100 strokes. Ask them to collect the fur in a sealable plastic bag. Ask the children to repeat this activity once a month. Attach the bags to a bulletin board. Ask the children to identify the month and the season when each sample is collected. Observe the color, texture, and quantity of hair collected during each month. Ask the children to guess why the animal's hair changes during the different seasons.

3. Distribute copies of the reproducible activity sheet on page 43. Follow the directions on the sheet.

Curriculum Crossovers:

1. Set up a bird feeder and observe the changing visitors. Which birds stay through the winter? Which birds are the first to arrive in the summer?

2. Discuss the different ways people get ready for winter.

3. Teach the children how to take their pulse. Compare their resting pulse rate and their pulse after active play. The pulse rate of a hibernating grizzly bear is 10-12 beats per minute.

Take a Trip:

Visit a garden or farm in the fall, winter, and spring. Record the flowers in bloom, the birds seen, young animals seen, where the animals are, the colors of the leaves.

Through the Year

The pictures show what a chipmunk does during each season. Cut out a picture and paste it in one of the boxes. Then draw a picture of something special you do in each season.

What a chipmunk does in the winter.	What I do in the winter.
What a chipmunk does in the spring.	What I do in the spring.
What a chipmunk does in the summer.	What I do in the summer.
What a chipmunk does in the fall.	What I do in the fall.

A Busy Year

by Leo Lionni (Alfred A. Knopf, 1992, F).

Willie and Winnie befriend a tree and watch it change during the four seasons.

Activities:

1. If your class did not "adopt" a tree with Mighty Tree you might adopt one now, (a deciduous tree is preferable) and follow it through the seasons. Draw pictures of the tree in September, January, April, and June. How does the tree change over the school year? (See further activities under TREES, pages 31-35.)

2. Stretch a metal hanger to form a square. Place it on a patch of grass or dirt under a tree. Collect examples of things found in the hanger square. You can dry samples from the squares between pieces of paper towel weighted down by books. After two weeks remove the samples and preserve them in clear contact paper.

3. Have children look for and record signs of the coming of each season. If necessary, prompt them by mentioning (not all at once!) insects, birds, other animals, flowers, trees, the sun, temperature, weather, etc. Encourage them to note sounds and smells, too.

4. Have children use the reproducible worksheet on page 45 to write haiku abut the changing seasons of the year.

Curriculum Crossovers:

1. Discuss the different things people do during the different seasons. How do people prepare for the coming of each season? What clothes do they wear? What activities do they do?

2. Make a collage of seasonal activities. This could include pictures from newspapers and magazines, family photos, etc.

3. Make a collection of seasonal clothes. Ask the children to sort the clothes by season. Ask them to explain why they sorted the clothes as they did.

Take a Trip:

1. Visit a nursery or greenhouse during each of the seasons to see what is for sale at that time and to observe how the workers prepare for each new season

2. Visit a nursery and learn about the special things trees need to grow healthy and strong. Compare and contrast these with the things people need.

From Season to Season

Each of these poems tells what a tree looks like during a different season. Write the name of the season in the box. Then draw a picture of the tree.

along the branches
tiny buds wait to open
and greet the sunshine

season:_____

the icicles hang
from the bare arms of a tree
like long, cold fingers

season:_____

dancing and dancing
among the tree's bright flowers
lovely butterflies

season:_____

shiny red apples
hang from the leafy branches
like colorful balls

season:_____

Write On! These poems are called *haiku*. Haiku are Japanese poems that often tell about things in nature. A haiku poem has three lines. The first line has 5 syllables, or beats. The second line has seven beats. The third line has five beats. What's your favorite season? Write a haiku that tells about it.

Air

- *MERLE THE HIGH FLYING SQUIRREL* by Bill Peet.

- *HOT-AIR HENRY BY MARY CALHOUN,* illustrated by Erick Ingraham.

Background:

1. Around Earth's surface, air occupies nearly every space between other kinds of matter.

2. Air takes up space.

3. Air exerts pressure.

4. Air rises when heated.

5. Wind is moving air.

6. Humans use air in many ways.

Literature Tie-Ins:

- **AIR BY DAVID BENNETT** (Bantam Little Rooster Books, 1989, NF)

- **THE AIR AROUND US** by Elenore Schmid (North-South Books, 1992, NF)

- **BIBI TAKES FLIGHT** by Michel Gay (Morrow Junior Books, 1984, F)

- **THE BIG BALLOON RACE** by Eleanor Coerr, illustrated by Carolyn Croll (Harper & Row, 1981, F)

- **FLYING** by Gail Gibbons (Holiday House, 1986, NF)

- **THE MAGIC BUBBLE TRIP** by Ingrid & Dieter Schubert (Kane/Miller Book Publishers, 1985, F)

Merle the High Flying Squirrel

by Bill Peet (Houghton Mifflin Co., 1974, F).

Merle, a timid Eastern city squirrel, decides to shed his fears and take a trip West to see the trees that are taller than buildings. When he takes time to untangle a kite, it carries him away. He has many adventures before landing in what he considers a "runt of a tree."

Activities:

1. Ask the children if air can move about and get into hidden spaces. Then place a brick in a clear container of water. Have the children observe and describe what they see happening.

2. Have children move their hand back and forth, like a fan, close to their face. Ask them to describe what they feel.

3. Give each child a piece of tissue paper. Have them hold their hand above their head, palm facing forward with the piece of tissue against it and walk forward. Stop suddenly. Why did the paper fall? What keeps the paper from falling when you walk?

4. Make an air-chair. Provide children with a garbage bag and a twist-tie. Have them drag the bag behind them until it is filled with air. Then have them twist-tie the opening and sit

on the bag. Ask: What are you sitting on? Have them describe how it feels. Encourage them to release some of the air and sit again. Ask them to compare how it feels with when it was full of air.

5. Provide each child with a plastic bag. Have them lay a book on top of the bag and gather together the opening of the bag so they can blow inside. Ask the children to explain what caused the book to move.

6. Use reproducible page 48 to help children make their own air-powered rocket ships.

Curriculum Crossovers:

1. Children can write a postcard to a friend describing what they saw while "flying" in a hot air balloon.

2. Brainstorm ways people use air. Make a bulletin board collage of things that use air.

Take a Trip/Invite a Guest:

1. Take a walk and look for evidence of moving air.

2. Contact one of the hot air balloon companies listed in the 800 directory, such as Sunrise Hot Air Balloons (1-800-548-9912) or Unicorn Balloon of Colorado (1-800-468-2477). Request information about hot air ballooning and a list of possible speakers.

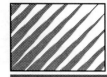

Air on the Move

You can make an air-powered rocket ship. Here is how.

● **What you need:**
10-foot piece of string • a straw • 1-quart milk carton •
scissors • long balloon • clothespin • tape • art materials

● **What you do:**
1. Tie one end of the string to a doorknob.
2. Slide the straw onto the string.
3. Cut out one long side of the milk carton. This open side will be the bottom of the rocket ship. Decorate the carton to look like a rocket ship.
4. Cut a small hole in the bottom of the milk carton. Put the balloon through the hole so that just the open end of the balloon sticks out.
5. Tape the top of the milk carton to the straw on the string.
6. Blow up the balloon. Stick the clothespin on the end of the balloon so that the air doesn't leak out.
7. Have a partner pull the loose end of the string tight. Make sure to hold it level with the height of the doorknob. Move the rocket ship to the end of the string your partner is holding. Take off the clothespin. What happens?

_____

What made your rocket ship move?_____

Hot-Air Henry

by Mary Calhoun, illustrated by Erick Ingraham (Mulberry Books, 1981, F).

Henry the cat wants to fly like everyone else in his family. When their guard is down, he makes a dash for the balloon basket, accidentally starts the burner and has his first solo flight.

Activities:

1. Using masking tape, attach two identical large brown paper bags to the ends of a stick, open ends down. Attach a string to the center of the stick and suspend it from a support (e.g. taped to the edge of your desk) so that the bags are 10-15 cm above the floor. Carefully balance the stick holding the bags. Then place a lit 100-watt bulb under one of the bags and observe what happens.

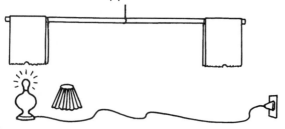

2. A parachute can be made from a plastic bag or a square of cloth such as a large handkerchief.

   a. Tie a piece of string to each of the four corners of the cloth or plastic.

   b. Attach the other ends of the string to a weight.

   c. Gather the parachute together and place the weight on the parachute.

   d. Throw the parachute up and watch it fall. If your school has two or more floors, drop the parachute down a stair well.

   e. Have the children explain what happened. Then ask: What do you think will happen if you add more weight to the object? (You can add clay or washers to the spool.) What do you think will happen if you increase the size of the parachute? What does the parachute catch as it falls?

3. Secure a gallon-size plastic bag over the opening of a wide mouth gallon jug with masking tape. Leave enough air in the bag so it stands up and make sure there are no air leaks around the tape. Ask students to try to push the bag into the jug. They will discover that air takes up space and exerts pressure.

4. Give the children balloons. Have them inflate their balloon and twist (but not tie) the end. Pose the question: What will happen if you let go of the balloon? Have them release the balloon and observe. Ask them to describe the journey their balloon took. Did all balloons travel in the same pattern? Why did some balloons travel farther than others?

5. Use the reproducible data sheet on page 50 to help students discover what happens when air is heated.

Curriculum Crossovers:

1. Hold an "Air Fair" featuring activities, pictures, and realia which use air. Exhibit kites, frisbees, parachutes, sailboats, etc. Invite parents and community members to participate by sharing their work or hobbies that relate to air.

2. Pretend you are Henry in the hot air balloon and you are sending radio messages to those on the ground.

Invite a Guest:

1. Invite an air hobbyist to talk to the class about the role that air plays in his or her hobby. Examples would be a person who flies kites, hang glides, or sky dives, who pilots, sails, or flies hot air balloons, or a meteorologist who knows about weather balloons. Use the 800 directory to look up numbers.

Up, Up and Away

What happens to air when it is heated? Try this simple experiment and find out.

 What you need:
glass bottle • balloon • a freezer

 What you do:
1. Put the bottle in the freezer for an hour.
2. Remove the bottle. Stretch the neck of the balloon over the bottle top.
3. Let the bottle sit for about 10 minutes. What happens to the balloon as the air inside the bottle warms up?

_____

_____

4. Warm air rises. How does this fact explain what happens to the balloon?

_____

_____

5. How do you think what happens to this balloon is like what happens to a hot air balloon?

_____

_____

_____

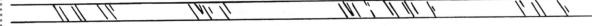

Hot and Cold: Which takes up more room: hot air or cold air? Here's another experiment you can try with balloons to find out. Fill two balloons with air so that they are the same size. Tie them closed. Put one balloon in the freezer for a day. Take it out of the freezer and put it next to the other balloon. Which is bigger? Let the cold balloon warm up again. What happens?

Weather

- *GILBERTO AND THE WIND* by Marie Hall Ets.
- *THE CLOUD BOOK* by Tomie dePaola.

Background:

1. Weather is caused by interactions of air, water, and the heat of the sun.

2. Heat from the sun causes air to expand and rise. Other air moves in to take its place, producing wind.

3. Water evaporates from the surface of the earth.

4. Evaporated water condenses to form clouds, which are made of tiny drops of water or snow.

5. The droplets of water in clouds can form larger drops and then precipitate as rain, snow, sleet, or hail.

Literature Tie-Ins:

- **CLOUDY WITH A CHANCE OF MEATBALLS** by Judi Barrett (Macmillan, 1978, F)

- **GERALDINE'S SNOW** by Holly Keller (Greenwillow, 1988, F)

- **LEGEND OF THE BLUEBONNET** retold and illustrated by Tomie dePaola (Scholastic, 1989, F)

- **RAIN** by David Bennett (Bantam Little Rooster Book, 1988, NF)

- **WEATHER FORECASTING** by Gail Gibbons (Macmillan, 1987, NF)

- **THUNDER CAKE** by Patricia Polacco (Philomel Books, 1990, F)

Gilberto and the Wind

by Marie Hall Ets (Puffin Books, 1963, 1978, F).

The story of a little boy and his playmate — the wind. The wind provides the boy with fun and frustration.

Activities:

1. Give each student an empty plastic produce bag. Go outside and have the children hold their bag open above their head and turn slowly until the bag fills with air. Children can feel the strength of the wind in the tug of a full bag versus the gentle pull of a less full bag of air. They can also see the strength of the wind as they watch the bag become full. Ask: What will happen if you let go? Try it and see.

2. Have the children use the reproducible worksheet to make a windsock (see page 53). Take the completed windsocks outdoors on a sunny day. First go to the sunny side of the building. Have children hold the string handles, letting their windsocks hang freely. Observe how they blow in the breeze (if at all). Now take the class to the shady side of the building. Again, observe the windsocks. Do they blow more in shade or in sun? What conclusions do the children draw from observing the different reactions of the windsocks in sun and shade?

3. Children can observe how heat affects air by placing a large pre-stretched balloon over the opening of a bottle that has been sitting at room temperature. Place the bottle with the balloon either outside in the snow or in a cooler of ice cubes. After about 30 minutes, bring the bottle back into the classroom. Have the children wrap their hands around the bottle. Ask them to describe what is

happening. Ask: Why do you think the balloon grew bigger?

4. Children can experience how wind helps to dry clothes by pouring equal amounts of water onto two identical handkerchiefs. Drape each over a clothes hanger and place them outside, one in a windy location and the other in a still location. Have the children check for dryness every 15 minutes. Ask the children to speculate why one dried faster than the other.

5. Blow bubbles and watch the wind carry them. For best results, do this when there is only a gentle wind or in an area protected from direct wind.

To make a quart of bubble mixture, pour 1/4 cup liquid dish detergent (Dawn or Joy is recommended; Ivory will not work) into a quart container. Add one tablespoon of glycerin (available at a pharmacy without a prescription, or from a science supply house). Add lukewarm water to fill, allowing it to slowly run down the side of the container to avoid bubbling over. For best results, allow the mixture to sit uncapped for 24 hours. It will then keep indefinitely when capped.

To make an inexpensive bubble pipe for each child, use a pencil to poke a hole in the side of a waxed cold drink cup, about 2 cm from the bottom. Cut straws in half, and place a half-straw into the cup hole. Invert the open cup into a bowl of bubble mixture. Pick it up and allow the excess to drip over the mixture. Hold the cup upright and blow out through the straw.

Experiment with different materials to see what size bubbles children can create. Which

direction do they travel? Do they all travel in the same direction? How high do they rise?

Children can blow a bubble inside a bubble by using a second half-straw which they have thoroughly moistened with the mixture. (Bubbles burst when they come in contact with a dry surface.) The children insert the wet straw into the first bubble and blow.

Have students observe the direction in which the bubbles move. Could there be a reason that they all move in one direction?

6. To demonstrate how hot and cold fluids interact, use two identical clear glass jars, such as pickle jars, and two smaller identical narrow mouthed glass jars, such as extract containers. The smaller jars do not have to be clear. Fill one large and one small jar with hot water. Fill the other two with cold water. Add a few drops of food coloring to each small jar. Use tongs to submerge the small jar with the hot water into the large jar with the cold water. Then submerge the small jar with the cold water into the large jar with the hot water. Ask the children to describe what they see. Have they ever seen something like this in nature?

Curriculum Crossovers:

1. Tell a story about real or imagined events on a windy day.

2. Brainstorm words to describe the movements and sounds of different kinds of winds.

Take a Trip/Invite a Guest:

1. Take a walk on a windy day. Ask the children to watch and remember what the wind carries or moves. Return to the classroom and ask: What things did the wind move? What things did the wind *not* move? Why did the wind move some things and not others?

2. Invite a meteorologist to talk to the class about weather forecasting.

Make a Wind Sock

 What you need:
a paper grocery bag • scissors • tape or glue • streamers or ribbon • yarn or string

 What you do:

1. Cut the bottom out of the grocery bag.

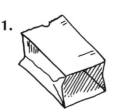

2. Fold down the top of the grocery bag.

3. Color the bag any way you like.

4. Staple streamers to the folded edges as shown.

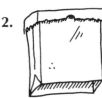

5. Staple string or yarn to the open end of the bag to make handles.

6. Hang from a place where it will blow freely in the wind and watch it go!

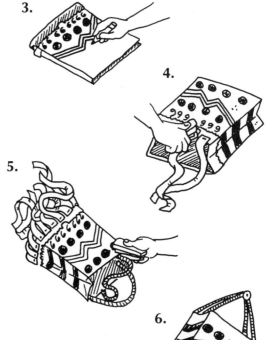

The Cloud Book

by Tomie de Paola (Holiday House, 1975, F).

All about the different types of clouds and the weather that follows them. The book also describes what people of different cultures think about clouds.

Activities:

1. You can make a cloud in the classroom with a one-gallon wide-mouth clear glass bottle (obtainable from a restaurant or delicatessen). Cut the neck off a balloon that is large enough to cover the opening of the bottle when used single thickness. (If necessary, you can secure it with a rubber band.)

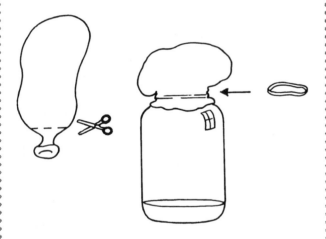

Place 5 cm of water in the bottle. Add smoke to the bottle from a smoldering match or two. Quickly cover the bottle opening with the balloon to contain the smoke, which should not be visible. With your hand, press the balloon into the bottle for 15 seconds, then catch the balloon with your fingers and thumb and pull up on it for 15 seconds. Repeat this 15-second pressing down and pulling up several times. With adequate smoke in the bottle a cloud should form several centimeters above the water. If not,

remove the balloon and add more smoke. Placing dark construction paper behind the bottle will make the cloud more visible.

2. Chill a glass, jar, or mirror for each child. Have children blow on it. Ask them to describe what occurs. The cloud is formed from tiny drops of warm moisture coming from the mouth and condensing on the cold surface. Ask the children if they have ever seen something similar happen in their homes.

Explain that this is called condensation and can frequently be seen on windows when the temperature outside is colder than inside, and when there is a lot of humidity in the air.

3. Using the reproducible worksheet (page 56), have the class observe and collect weather data daily. Include the type of clouds, temperature, precipitation, wind, and other appropriate data. Record the data on the worksheet weather log.

4. Set up a water cycle. In a clear glass or plastic container, put a small amount of water to which food coloring has been added. Cover with plastic wrap and secure with a rubber band. Set the container in sunlight. Ask the children to discuss what they observe.

5. Collect raindrops in a pie pan. Put a 2 cm deep layer of flour in the bottom of the pan. Place it outside at the beginning of a rainstorm to catch the big raindrops. Bring it in as soon as a few drops are collected (or else all of the flour will become soggy). Allow it to dry

undisturbed. Then sift out the lumps formed by the raindrops. Are all these raindrops the same size? Repeat the activity during different rainstorms and compare the size of the drops obtained.

6. Collect rainfall for a month or more. Use a straight sided open jar. Measure the amount of rain collected after each rainfall. Record the data and use it to construct a graph. Compare the rainfall for a period of time. Ask the students questions that focus on differences and similarities in rainfall.

Curriculum Crossovers:

1. Write a story about the adventures of a raindrop.

2. Have the children recite or sing any songs they know about rain.

Take a Trip/Invite a Guest:

1. Take a walk on a sunny day when there are lots of clouds in the sky. Questions to ponder: Are all the clouds going the same direction? Do they always travel in that direction? Are they all traveling at the same speed? Have any of them formed familiar shapes?

2. Ask someone who flies frequently to come to class to tell of his or her experiences flying through heavy clouds.

Name

Watching the Weather

What's the weather like today? Use this chart to keep track of the weather. Then answer the questions below.

Date	Temperature	Rain or Snow	Wind	Clouds

1. Which day was the warmest?_____

2. Which day was the coolest?_____

3. Did any rain or snow fall this week?_____ If yes, on which day(s)?_____

_____

4. Why do people want to know what the weather will be like?_____

_____

5. What jobs are affected by the weather?_____

_____

Eyes, Noses, Feet

- *EYES* by Judith Worthy.
- *BREATHTAKING NOSES* by Hana Machotka.
- *BOOT WEATHER* by Judith Vigna.

Background:

Eyes:

1. Eyes are used for seeing.
2. Nearly all larger animals — fishes, reptiles, birds, mammals — have two eyes that allow them a wide field of vision.

Noses:

1. Most noses are used for breathing. Some noses are used for smelling. Some animals use their noses for digging or grasping things.
2. Animal noses are adapted for the environment where the animal lives.

Feet:

1. Feet are the parts of the body on which humans and animals stand and move.
2. Feet come in many shapes and sizes. They differ according to each animal's particular needs and environment.
3. Most human feet have five toes, an arch, a sole, and a heel.

Literature Tie-Ins:

- **WHAT DO YOU DO AT A PETTING ZOO?** by Hana Machotka (Morrow Junior Books, 1990, NF)
- **WHAT NEAT FEET** by Hana Machotka (Morrow Junior Books, 1991, NF)
- **ALL KINDS OF FEET** by Ron and Nancy Goor (Crowell Jr. Books, 1984, NF)

- **MOCKINGBIRD MORNING** by Joanne Ryder (Four Winds Press, 1989, F)
- **STEP INTO THE NIGHT** by Joanne Ryder (Four Winds Press, 1988, 32 pp., F)
- **MY FEET** by Aliki (Thomas Y. Crowell, 1990, NF)
- **UNDER YOUR FEET** by Joanne Ryder (Macmillan, 1990, F)
- **AN ELEPHANT NEVER FORGETS ITS SNORKEL** by Lisa Gollin Evans (Crown Publishers Inc., 1992, NF)

 Eyes

by Judith Worthy (Doubleday, 1988, NF).

A colorful look at the eyes of many different creatures.

Activities:

1. Children can use small mirrors to observe their eyes. (Tape the mirror edges for safety.) Darken the room for one minute and then turn all the lights on. The children can use the mirrors to observe the dilation of their pupils.
2. Children can draw and color a picture of their own eyes. After they have finished their drawings the children can group themselves by eye color. First ask them to predict which group they think will be the biggest. Record their guesses. Now the children can create a living "graph" of their eye colors. After they have grouped themselves, ask each group to form a single line. Arrange the lines in order from shortest to longest. This is their living graph. Tape the pictures to the blackboard, arranging them in the same order as the student lines.

3. Children might enjoy surveying another class to determine the eye colors of those students. They can also graph this information.

4. Have the children use the reproducible worksheet on page 59 to match the description of the eyes to the animal the eyes belong to.

Curriculum Crossovers:

1. Children can cut out pictures of eyes showing different emotions. They can select one picture and write a story about those eyes. Who do they belong to? What is the person feeling? What happened to make him or her feel this way?

2. Children can cut out pictures of eyes and use them as the basis for creating new animals. Ask the children to describe what their creature would see.

3. Children can write to the American Foundation for the Blind at 15 West 16th Street, New York, NY 10011 and request information about the training of guide dogs.

Take a Trip/Invite a Guest:

1. Visit an eye center or an optometrist's office for a demonstration of eye testing equipment.

2. Ask a person with a seeing eye dog to visit the class.

Look at Those Eyes!

Each riddle tells about one of the animals on the right side of the page. Read each riddle. Look at the pictures. Then write the name of the animal that answers each riddle.

1. I have two eyes at the front of my head. My eyes have very large pupils to help me see at night.

What am I?_____

snail

2. I have an eye on each side of my head. This helps me watch out for enemies coming from different directions.

What am I?_____

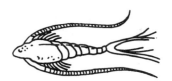

copepod

3. I have just one eye at the top of my head. It can only make out light, gray shapes. Two sets of antennae help me feel what I cannot see.

What am I?_____

owl

4. I have two long tentacles that stick up from the top of my head. My eyes are at the end of these tentacles. I can bend the tentacles to look for danger in all directions.

What am I? _____

chameleon

Write a Riddle: Find out about the eyes of another animal. Write a riddle about how it sees. Share your riddle with a friend.

Breathtaking Noses

by Hana Machotka (Morrow Junior Books, 1992, NF).

A book about specialized noses in the animal world.

Activities:

1. Place pieces of cotton dipped in various scents (coffee, lemon juice, extracts, etc.) in containers such as empty 35mm film canisters. Ask the children to match each container to the items or pictures of the items that made each scent.

2. Some animals recognize each other by their scents. Again place pieces of cotton dipped in various scents in containers but this time make two of each container. Arrange the students in a group in the center of the classroom. Give one container to each student. Tell the students to find their partner using only their sense of smell.

3. Create three different scent trails in the room by placing the scents on construction paper. Each piece of construction paper should be the same size, shape, and color. Rub or sprinkle each scent on the pieces of construction paper that will make up its trail. The trails can be as long or as short as is developmentally appropriate for the children, and they can overlap. Divide the class into three groups. They must find their way through their maze only with their noses.

4. Have children use the reproducible worksheet on page 61 to write a nose fable.

Curriculum Crossovers:

1. Collect pictures of different animal noses. Children draw in the rest of the animal and write a story about it.

2. Ask the children to pretend they have the nose of an elephant (or pig, snake, etc.). They can write a story about what it is like to have that nose.

3. Collect play animal noses and compare them to each animal's real nose.

Take a Trip:

Visit a zoo or pet store to observe noses. How do other animals use their noses?

Name _____

The Story of a Nose

There's an old story that tells how the elephant got its long nose. The story says that one day a crocodile tried to eat a baby elephant. The crocodile grabbed hold of the elephant's nose and pulled and pulled, until its nose stretched out long.

Look at the interesting noses on these animals. Select one of the animals. Then work with friends to make up a story about how it got its unusual nose.

_____

_____

_____

_____

_____

_____

_____

_____

_____

_____

_____

_____

_____

_____

_____

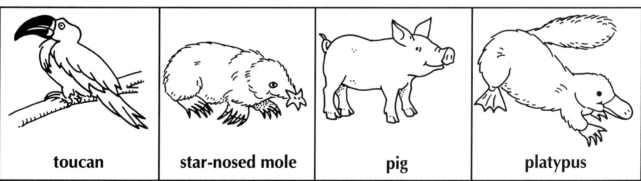

| toucan | star-nosed mole | pig | platypus |

Know More About Noses: How does the animal you wrote about use its nose? Use books to help you find out.

Boot Weather

by Judith Vigna (Albert Whitman & Co., 1989, F).

While playing in her backyard, a little girl imagines she is many other people wearing many different kinds of boots.

Activities:

1. Sit in a circle and ask each child to remove one shoe or other footwear. Have them place their footwear in front of them. Ask questions that force them to observe closely, such as: Are there any shoes that are alike? How can the shoes be grouped? Ideas for grouping include by types of material, types of fasteners, color, height, number of eyelets, length of laces, etc. Each time the students suggest a new characteristic have them regroup the footwear on the floor.

2. The children can make shoe and boot prints. Have the children remove one shoe, lay it sole side up, and wrap it in a piece of newsprint. It is best for children to hold the shoe between their knees. Have them rub across the sole with the flat side of a crayon. You can cut out the rubbings to leave just the outline of the shoe or boot. Display the rubbings and ask the children how they can be grouped. Ask the children to compare the prints and arrange them from largest to smallest, longest to shortest, widest to thinnest.

3. Have the children draw footwear appropriate to different people and their jobs, e.g., firefighters, ballerinas.

Curriculum Crossovers:

1. Children can demonstrate how to run, skip, march, kick, hop, walk on tiptoe, skate, and dance.

2. Children can write a story about who or what is underfoot when they walk.

3. Have the children bring in old footwear. They can then:

   a. choose any pair, assume the role of the person who could have worn it, and act out a skit.

   b. arrange the footwear from heaviest to lightest.

   c. arrange the footwear from tallest to shortest.

   d. arrange from the one that could hold the most to the one that could hold the least. Use styrofoam packing pellets to determine the volume.

   Extend this activity into the home by having students search for the most unusual shoe sole pattern or the most unusual fasteners.

5. Brainstorm all the things that are worn on feet.

6. Start a "shoe shop" in your classroom. Borrow a measuring device and perhaps the stool to sit on when measuring.

7. Cut out pictures of footwear. Display them and ask the children how they can be grouped.

Take a Trip:

Take your class on a track hunt. You can make plaster casts of animal tracks. First place a "fence" around the track by using a strong strip of oak tag or a milk carton with the top and bottom cut off. Pour a plaster of paris mixture over the track and allow it to harden. Carefully remove the cast when it is dry. Use a soft toothbrush to clean off unwanted dirt, etc. It may be necessary to rinse the tracks in water.

Track Match

These cards show pictures of animals and the tracks their feet make. Which animal's feet make which track? Cut out the cards and turn them face down on a table. Use the cards to play a game of lotto with a friend. Whoever gets the most correct matches wins.

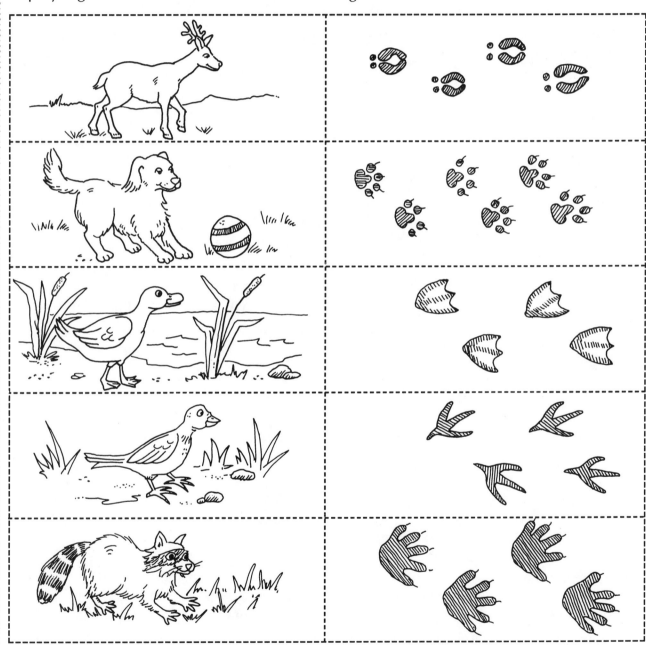

Make a Track Trap: Put some seeds, nuts, pet food, or a piece of an apple on a paper plate. Find a place outside near where animals live. Put the plate on the ground. Spread some mud around the plate (you can mix some in a jar with dirt and water). Then check on your plate the next day. Do you see any tracks in the mud? What animals do you think made them?

- *THE VERY QUIET CRICKET* by Eric Carle.

- *SNAIL'S SPELL* by Joanne Ryder.

Background:

Like all animals, insects, spiders, and snails have adaptations — different body parts and behaviors — that enable them to survive in their different environments.

Insects:

1. Insects have three body parts and six legs. Many insects have wings, and most have antennae.

2. Some insects go through a complete metamorphosis from egg to larva, to pupa, to adult.

3. Other insects go through a simple metamorphosis from egg to nymph, to adult.

Snails:

1. Snails are mollusks, having soft bodies and no bones, and shells, like clams and oysters.

2. Land snails can live two to three years.

3. Snails live in a moist environment.

4. The under part of a snail is a large muscular surface called a foot because the animal uses it to move. Movement is aided by a kind of slime produced at the front of the foot.

5. Snails have an operculum or horny plate that tightly closes the opening in its shell after it retreats.

Literature Tie-Ins:

- **THE ICKY BUG ALPHABET BOOK** by Jerry Pallotta (Charlesbridge, 1986, NF)

- **THE SPIT BUG WHO COULDN'T SPIT** by Penny Pollock (Putnam Publishing Group, 1982, F)

- **SNAIL AND CATERPILLAR** by Helen Piers (American Heritage, 1972, F)

- **THE BIGGEST HOUSE IN THE WORLD** by Leo Lionni (Alfred A. Knopf, 1968, F)

- **BE NICE TO SPIDERS** by Margaret Bloy Graham (Harper & Row, 1967, F)

- **SPIDERS IN THE FRUIT CELLAR** by Barbara Joosse (Alfred A. Knopf, 1983, F)

- **LEESE WEBSTER** by Ursula K. LeGuin (Atheneum Books, 1979, F)

- **THE LADY AND THE SPIDER** by Faith McNulty (Harper & Row, 1986, F)

The Very Quiet Cricket

by Eric Carle (Philomel Books, 1990, F).

A young cricket meets many other insect neighbors but is unable to give them a greeting until he meets a female cricket.

Activcities:

1. Collect insects, including crickets, for classroom observation. (You can purchase crickets, and mealworms for feeding them, at pet stores.) Bug boxes or other magnifiers

increase the excitement of looking at insects. Observe how different insects move, fly, and eat. Do they make any kind of noise?

2. Observe crickets. What color are they? What is the shape of their wings? Can you find their mouth parts? What do they eat? When do they eat? What do their legs look like? How many legs do they have? Are all of their legs the same size? How do they move? Do they have antennae? What do their eyes look like? Have the children draw their crickets.

3. Listen to the chirps of crickets. Does the frequency of chirps differ in warm and cold weather? Can the children imitate the sound of the cricket?

Curriculum Crossovers:

1. Do a comparative creative writing lesson with the following verbs from Very *Quiet Cricket:*

   whizzed like a...

   whispered like a ...

   crunched like a ...

   bubbled like a...

   screeched like a ...

   hummed like a ...

2. Have the children create insects from the reproducible worksheet (page 66) and then write stories about their creation. Where does it live? What are its special adaptations?

3. You can have the children observe night insects by placing a bright light near a glass window. Insects will collect on the outside of the window and can be observed up close with a magnifier. Placing the light near a screened window will allow the children to hear the sounds, if any, that the insects make but will make clear viewing a little more difficult. When the children return to class ask them to describe the most unusual insect they observed.

Invite a Guest:

1. Invite a local reporter to describe to students what a reporter does. Afterward the children can pretend to be reporters interviewing an insect.

2. Invite an insect collector to talk about gathering and labeling an insect collection. Ask him or her to describe some of the ways insects help people and some of the ways they harm people.

Make-A-Bug

Directions: Cut out assorted body parts to create your own insect. Include three body parts, antennae and six legs.

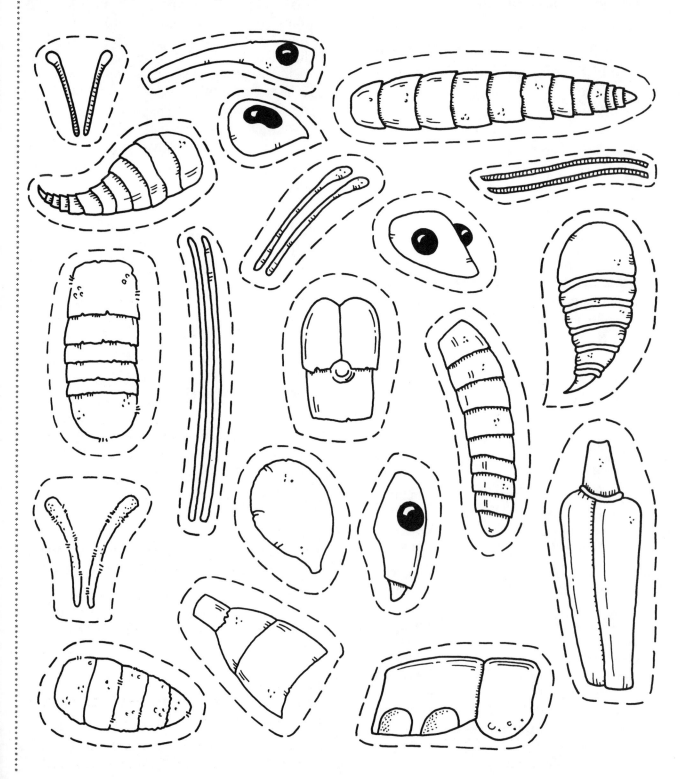

Snail's Spell

by Joanne Ryder (Puffin Books, 1988, F).

As a small boy imagines what it would be like to be a snail, he shrinks and begins to experience life as a snail. A close up look at a snail's life and habitat.

Activities:

1. Snails have eyes at the ends of a pair of tentacles and do not see as clearly as people. You can simulate the eyesight of a snail by making a snail hood. Cut an opening 10 cm high by 20 cm wide near the top of a brown paper bag. Cover the opening with a piece of wax paper. Ask the children to place the bag over their head with the wax paper opening in front of their eyes. Ask the children to describe what they see. How do they think a snail sees the world?

2. Children will enjoy observing the movement of a snail. Obtain a snail from a local aquarium or pet shop. Place the snail on a transparent surface such as a clear plastic tumbler or the plastic lid of a card box and have the children observe its movement from below. Ask the children to describe how a snail moves. Could they move as rapidly in the same way? What other animals can they think of that can move across surfaces vertically or upside down?

3. Other questions to stimulate observation and discussion include: Does a snail make any noise? What does a snail do when it is frightened? How do snails eat? What do they eat? Do they have preferences in food? How much do snails eat? Do snails eat more at one time of the day than another?

4. Snails have a tongue-like structure called a radula. A snail's radula can have as many as 14,000 teeth on it. Have the children examine their own tongues with a small mirror and a hand lens. Have the children examine the surface of sandpaper with a magnifier. Have the children rub their fingers over the sandpaper and describe what it looks and feels like. The surface of sandpaper is much like the surface of a snail's tongue.

5. Using the reproducible worksheet on page 68, have children sort the minibeasts.

Curriculum Crossovers:

1. Describe what it would be like to be a snail.

2. Put on the snail hood and observe some common things you see every day. Write a description of these things through the eyes of a snail.

Take a Trip:

1. Go outside and try to find snails.

2. Visit a pet store and observe water snails. How do water snails differ from land snails?

Sorting Minibeasts

Read the fact boxes about two kinds of minibeasts: insects and arachnids. Then cut apart the pictures at the bottom of the page. Paste the arachnids in the left column. Paste the insects in the right column.

Arachnid Fact Box	**Insect Fact Box**
• have four pairs of legs • don't have wings • bodies are divided into two parts	• have three pairs of legs • many have two pairs of wings • bodies are divided into three parts

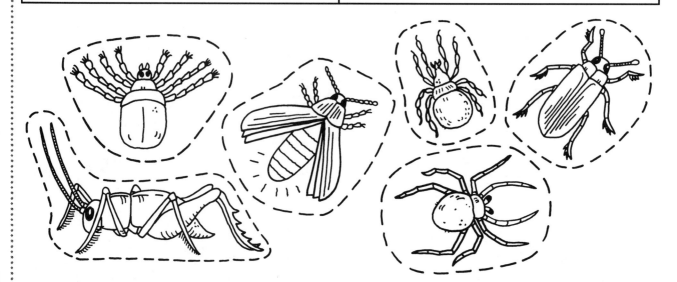

More Minibeasts: A snail is a mollusk. Make a fact box that tells about mollusks. Find out the names of two other kinds of mollusks.

Animal Homes

- *IS THIS A HOUSE FOR HERMIT CRAB?* by Megan McDonald.

- *MOUSEKIN'S GOLDEN HOUSE* by Edna Miller.

Background:

1. Animals live almost every place on the Earth in many different kinds of homes.

2. Animals select homes that provide protection from weather and predators.

3. Most animals make their own homes, using materials available in their environment.

4. Some animals grow their "homes" in the form of shells. Hermit crabs, however, cannot grow shells. Instead, they use the discarded shells of mollusks, changing shells as they grow larger. Other kinds of crabs have an exoskeleton (hard outer skin like a shell), which they shed as they grow larger. Their inner skin then hardens to replace it.

Literature Tie-Ins:

- **THE ARCHITECTURE OF ANIMALS** by Adrian Forsyth (Camden House, 1989, NF)

- **HOUSE FOR HERMIT CRAB** by Eric Carle (Picture Book Studio, 1987, F)

- **APPLEMOUSE,** translated by Timothy Cleary (Hill and Wang, 1984, F)

- **THE BIGGEST HOUSE IN THE WORLD** by Leo Lionni (Alfred A. Knopf, 1968, F)

- **BUSY BEAVERS** by Lydia Dabcovich (Scholastic, 1988, F)

 ## Is This a House for Hermit Crab?

by Megan McDonald (Orchard Books, 1990, F).

The trials and travels of a hermit Crab

searching for a new shell after he has outgrown his old one.

Activities:

1. Discuss houses with the children. What makes something a house? Why is a house necessary? Are all houses alike? What can homes be made out of?

2. Compare and contrast different animal homes. What materials do they use? (Shells, wood, leaves, twigs, dirt, mud?) Are their building materials natural or human-made? How often do they have to make a new home? Do they have more than one home?

3. Observe a hermit crab purchased from a pet store. Since you cannot return it to its habitat, have the children keep and feed it as a pet.) How many body parts does it have? How does it move? What does it eat? What does it do when it is frightened? How does it find its food? Can you find its eyes, ears, nose, mouth?

4. Using the reproducible worksheet on page 70, have the children decide which habitats are suitable for which animals.

Curriculum Crossovers:

1. Compare and contrast houses in your community, in other parts of the U.S., and with other homes around the world. The children may want to collect pictures from magazines for this activity.

2. Compare and contrast different shells. Would all shells make good hermit crab homes? Which are best?

Take a Trip/Invite a Guest:

1. Discuss the types of homes people provide for animals. Have a naturalist bring a bat house into the classroom.

Name

Animals At Home

Circle yes or no after each question. Then draw pictures of the homes for the animals in the last two boxes. Use books to help you.

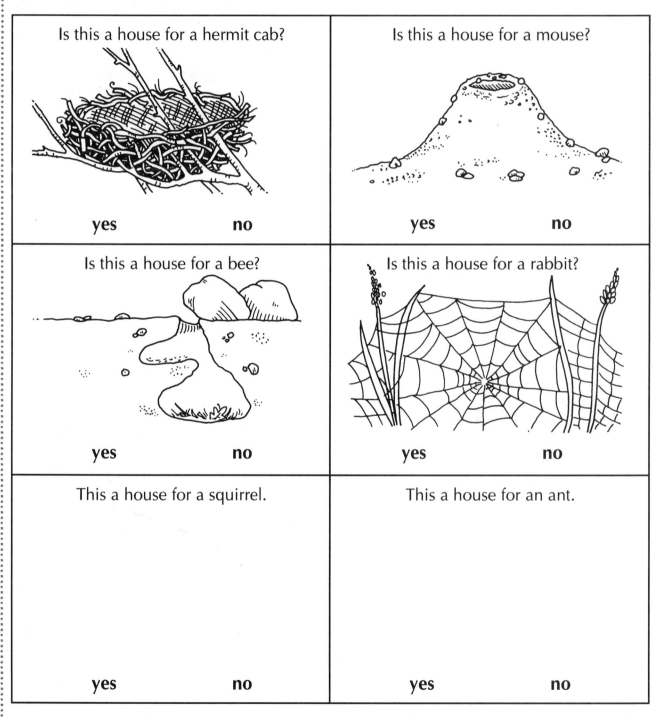

Is this a house for a hermit cab?

yes   no

Is this a house for a mouse?

yes   no

Is this a house for a bee?

yes   no

Is this a house for a rabbit?

yes   no

This a house for a squirrel.

yes   no

This a house for an ant.

yes   no

House Hunt: For the questions you answered "no" to, draw a picture on the back of this page of what the animal's house looks like. Then pick one of the animals (except the hermit crab!) and write a story about its search for a home.

Mousekin's Golden House

by Edna Miller (Prentice-Hall Books for Young Readers, 1964, F).

A mouse discovers a discarded jack-o'-lantern, which he turns into his home.

Activities:

1. Place a jack-o'-lantern in front of the children and ask: Who could live in this house?

2. Children can observe a jack-o'-lantern decaying in a large fish tank. Place soil in the bottom of the tank to make a layer 5 cm deep. Place the jack-o'-lantern on the soil. Carefully seal the tank with clear contact paper or heavy-duty plastic wrap cut 5 cm larger on all sides of the opening. Observe the gradual decay of the jack-o'-lantern and also the water cycle that accompanies it. Have children predict how long it will take the jack-o'-lantern to decay down to the level of the dirt.

3. Keep a decaying pumpkin picture diary (see the reproducible worksheet on page 72).

Curriculum Crossovers:

1. Compare different types of pumpkins (see the Activities under Seeds and Plants, page 26).

2. Tell a story about a mouse that uses a different object for a home.

3. Brainstorm things that are the same color as a pumpkin.

Take a Trip:

1. In the spring, go to a seed store and find out how many types of pumpkin seeds are sold.

2. In the fall, go to the market and find out how many kinds of pumpkins are sold.

Name _____

Pumpkin Diary

Directions: Draw a picture of how the pumpkin looks at the end of each week. Write a sentence to tell about each picture you draw.

week 1	week 2

week 3	week 4

week 5	week 6

Birds

Background:

1. Birds are warm blooded vertebrates with two legs and two wings, a beak and no teeth. Their bodies are covered with feathers.

2. Birds range in size from tiny hummingbirds to ostriches taller and heavier than humans. Birds of one kind or another are found in nearly all regions of Earth.

3. The bones of birds are hollow and very light. Most birds can fly.

4. Birds eat a lot of food because flying uses up a lot of energy and they also need to maintain their body temperature. Many smaller birds eat their weight in food every day.

5. Female birds lay eggs that have a protective shell.

Literature Tie-Ins:

- **BABY NIGHT OWL** by Leslie McQuier (Random House, 1989, F)

- **GOOD-NIGHT OWL!** by Pat Hutchins (Macmillan, 1972, F)

- **JUST PLAIN FANCY** by Patricia Polacco (Bantam Books, 1990, F)

- **MOCKINGBIRD MORNING** by Joanne Ryder, illustrated by Dennis Nolan (Four Winds Press, 1989, F)

- **OWL MOON** by Jane Yolen, illustrated by John Schoenherr (Philomel Books, 1987, F)

- **WHOO-OO IS IT?** by Megan McDonald, illustrated by S.D. Schindler (Orchard Books, 1992, F)

Make Way for Ducklings

by Robert McCloskey (Puffin Books, 1941, 1969, F).

A pair of mallard ducks migrates north looking for a place to build a nest and raise a family. Finally a proper location is found, the nest is built, and eggs are laid. The young ducklings survive life in the city with the help of a friendly policeman.

Activities:

1. Construct two different types of beaks out of popsicle sticks and spring-type clothespins (two of each for each beak):

   a. Measure off one fourth of the length of each stick, and cut the sticks in two at that point.

   b. Put a small piece of waxed paper between the two smaller popsicle pieces.

   c. Open one clothes pin and put a dab of glue on each of the exposed inner surfaces.

   d. Place the cut ends of the small popsicle sticks onto the dabs of glue and close the clothes pin.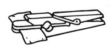

   e. Put a small piece of waxed paper between the two larger popsicle pieces.

f. Open the other clothes pin and put dabs of glue on the inner surfaces.

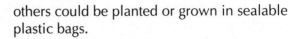

g. Place the cut ends of the large popsicle sticks onto the dabs of glue and close the clothes pin.

h. Leave the "beaks" overnight for the glue to set. Remove the pieces of waxed paper.

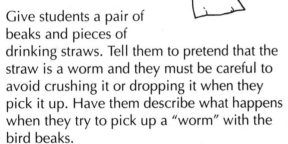

Give students a pair of beaks and pieces of drinking straws. Tell them to pretend that the straw is a worm and they must be careful to avoid crushing it or dropping it when they pick it up. Have them describe what happens when they try to pick up a "worm" with the bird beaks.

Compare the bird beaks you constructed with the drawings on the reproducible worksheet (page 75). How do they compare with the beaks of real birds?

2. Set up a bird feeder filled with wild bird seed mixture outside the classroom. Keep records on which birds come to the feeder and what they prefer to eat. Observe how the birds eat the seeds. Questions to answer: Do the birds prefer some seeds to others? Do all birds eat the same seeds? Do some birds take seeds away from the feeder to eat? Are some birds neater eaters than others?

After several weeks try only sunflower seeds in the feeder. Do different birds come to the feeder?

3. Have the children examine wild bird seed with a magnifying glass. Ask: Do you think each piece is a seed? Why or why not? How could you prove it was or wasn't a real seed? Try growing the bird seed in the classroom. Some could be placed on a damp sponge,

others could be planted or grown in sealable plastic bags.

4. Use the reproducible worksheet on page 74 to help students make a mini-book about bird beaks.

Curriculum Crossover:

1. Borrow a bird song record or tape from the library or a parent and play it for the children. Ask them if they have heard birds sing similar songs. Are all bird songs alike? How do they differ?

2. Ask students to describe what it would be like to be the last duckling in line when they crossed Beacon Street.

Take a Trip/Invite a Guest:

1. Invite a bird watcher to come to the class, if possible with slides to show. Ask him or her to explain the many ways a bird can be identified.

2. In the spring, take a walk to listen for birds singing. How many different songs can you hear?

A Peak at Beaks

Different kinds of birds have different kinds of beaks. The birds use their beaks to eat different kinds of foods. Make a mini-book about bird beaks. Just follow the directions below.

 What you need:
scissors • crayons • paste • books about birds

 What you do:
1. Cut apart the pages on page 76.
2. Staple the pages together in order, with the cover on top. (Staple them together on the left-hand side of the pages.)
3. Color the pictures of the birds below. Cut them out.
4. Read each of the pages that tells about a bird's beak. Try to find the bird with the beak the page tells about.
5. Paste the picture of the bird to the opposite page that tells about it. (Use books to help you check if you are right before you paste the pictures in your book.)
6. Make up a title for your bird beak book. Write it on the cover. Then draw a picture on the cover of your book.

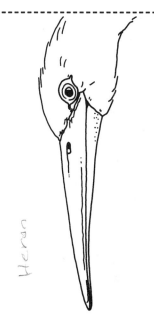

Heron

Kite

Woodpecker

duck

Pelican

A Peak at Beaks (cont.)

(**Note**: The blank page will be the cover of your book.)

Kites eat snails. They use their slender, curved beak to reach inside the snail's shell and pull the snail out.

1

Woodpeckers eat insects. They use their pointy beak like a chisel to drill holes into trees and look for insects.

2

Ducks eat plants that float on the water. They use their wide, flat beak to skim the plants.

3

Pelicans have a stretchy throat pouch under their beak. This helps them scoop up the fish they eat.

4

Herons have a long, pointed bill that they use like a spear to catch fish.

5

Owlbert

by Nicholas Harris (Garth Stevens Children's Books, 1989, F).

Nicholas wants a pet but his parents disapprove of his choices. He keeps it a secret when he discovers Owlbert. His parents are astounded when an owl joins them for a barbecue.

Activities:

1. Use the reproducible worksheet (page 78) to study the different types of bird feet. Birds use their feet to walk, perch, swim, run, climb, and grab. Can students tell which feet perform which activity best?

2. Ask the children to design a new bird with the help of this reproducible and the one from *Make Way For Ducklings*. When they have finished, have them tell where their bird lives and what it eats.

3. Owls come in many sizes. Find out about the largest and smallest owls in your area. Measure the sizes and compare them to common birds found at feeders.

4. Obtain owl pellets (the nondigestible mass of bones, teeth, hair, and feathers regurgitated by owls) from science supply houses or from Creative Dimensions, P.O. Box 1393, Bellingham, WA 98227.

   Provide groups with a paper plate, an owl pellet, and a sharp pencil (or tweezers or a sharpened piece of doweling if available).

   Instruct students to gently pull the hair and feathers away from the bones, using the tools provided, and carefully lay the bones aside. Questions to focus on: Are all the bones alike? Do the bones look as if they come from the same animal? Could you try to reconstruct an animal from the pellet contents?

Curriculum Crossovers:

1. Ask students to identify what they think the ideal pet would be. Have them write about how they would care for their pet.

2. Have students heard the expressions "wise as an owl" or "a wise old owl"? Why do they think owls acquired a reputation for wisdom? What other birds are often given human characteristics?

3. What word is used to describe the call of an owl? Brainstorm words for other bird calls (such as quack, caw, tweet).

Invite a Guest:

Invite a naturalist or rehabilitator to the class to talk or bring an owl to school. Have him or her discuss the real problems a person would encounter if they had to care for an owlet.

Which Feet Fit?

These birds all have different feet. But the feet are on the wrong birds. Follow the directions to find the right feet for each bird.

1. Cut apart the strips on the dotted lines.
2. Stack them on top of each other.
3. Staple them along the left-hand side as shown.
4. Cut along the solid lines at the bottom of all the strips. Be careful not to cut all the way to the end of the paper. Stop where the line stops.
5. Read the lines that tell about each bird's feet. Then flip the bottom part of the book back and forth until you find the right feet to fit each bird.

Duck

I use my webbed feet to help me swim. They work like paddles.

Osprey

I use my large, curved claws to grab fish from the water.

Robin

I have a long hind toe which helps me grab onto a branch tightly.

Ostrich

I have only two toes. By running on my tiptoes, I can move very fast.

Whales

- *I WONDER IF I'LL SEE A WHALE* by Frances Ward Weller, illustrated by Ted Lewin.

- *DEAR MR. BLUEBERRY* by Simon James.

Background:

1. Whales are mammals that live in the ocean. Whales can be grouped into those that have teeth and a single blowhole, and those that have two blowholes. They can also be grouped as either "toothed" or "baleen".

2. There are many different species of whales, differing in size and shape. The Blue Whale is the largest animal to have ever lived. It can be 100 feet in length and weigh 120 tons.

3. Like other mammals, baby whales drink their mother's milk, which is rich in vitamins. A baby gray whale may drink 50 gallons of milk a day and gain 200 pounds daily.

4. Whales have special mouth parts for obtaining food. Some have baleen (strips of a hornlike substance), which acts like a strainer separating the food from the water. Others have teeth to catch and rip their food, which may consist of plankton, krill, or fish.

   Plankton are microscopic plant and animal organisms floating in water in large numbers. Krill are small crustaceans. Baleen whales usually do not go as deep as toothed whales because their food is found near the surface. A baleen whale can gulp or skim krill from the water. Humpbacks are baleen whales. Although they grow over 15 meters (50 ft) in length, they cannot swallow any piece of food larger than a baseball. They spend three or four summer months off Alaska. During that time they consume a full year's supply of food.

5. Each whale species has its own variety of sound or song, which can be a click, groan, whistle, creak, burp, cry, scream, or chirp.

Some sounds are inaudible to humans but can be heard by other whales thousands of miles away.

6. Whales migrate to warmer waters in winter.

7. Individual whales are identified by the markings on their flukes (tail fins).

Literature Tie-Ins:

- **BABY BELUGA** by Raffi (Crown Publishers, 1980, 1992, F)

- **BURT DOW** by Robert McCloskey (Viking Press, 1963, F)

- **HUMPHREY THE WAYWARD WHALE** by Ernest Callenbach, illustrated by Christine Leefeldt (Heyday Books, 1986, F)

- **OCEAN ANIMALS, A RANDOM HOUSE TELL ME ABOUT BOOK** by Michael Chinery, illustrated by Eric Robson (Random House, 1992, NF)

- **WHALES** by Gilda Berger (Doubleday, 1987, NF)

- **HOW THE WHALE GOT HIS THROAT** by Rudyard Kipling (Philomel Books, 1987, F)

 ## I Wonder if I'll See a Whale

by Frances Ward Weller, illustrated by Ted Lewin (Philomel Books, 1991, F).

While on a whale watch a young girl observes a mother whale with its baby swim close by the boat before disappearing into the deep. As the day goes on, she sees a whale breach and watches terns and gulls take food from the mouths of humpbacks.

Activities:

1. In the hallway or at the playground, help students measure the average length of selected whales:

Pilot whale	**20 feet**
Gray whale	**40 feet**
Humpback whale	**50 feet**
Sperm whale	**55 feet**
Right whale	**55 feet**
Finback whale	**70 feet**
Blue whale	**100 feet**

Compare the whale lengths to the size of the classroom, gym, lunchroom, hall, etc., or to the length of a car, truck, bus, etc.

2. Simulate whales feeding by adding 5 cm of water to a wide shallow container. Cut short lengths of grass onto the water to represent krill and add cut circles of raw carrot to represent small fish.

Provide some students with tongs to simulate the teeth of a toothed whale and others a comb to simulate the baleen of a baleen whale. They must hold the comb in a vertical position only. Have the children pretend they are hungry whales and see how much food they can catch.

Questions to pose: Would a baleen whale be very successful in catching krill? How successful would a toothed whale be at catching krill?

Remind the children that, in the real world, plankton, krill, and fish are not passive. Also, the baleen whale takes in a lot of water which must be pressed out before it swallows the krill.

3. Have children use the reproducible pattern on page 81, and follow the directions to create a whale puppet.

Curriculum Crossovers:

1. On a map, identify the migration routes of whales.

2. Have the children try to imagine what it would be like to be able only to strain food and swallow it without chewing. Ask them to describe how they think it would feel.

3. Lead the class in a game of Simon Says, substituting the terms *starboard* for right and *port* for left. These may be unfamiliar words to the children but are the norm for sailors.

4. Adopt a whale. Write to: International Wildlife Coalition, 634 North Falmouth Highway, P.O. Box 388, North Falmouth, MA 02556 for information on the junior members' "Whale Adoption Project."

Invite a Guest:

1. Invite a guest to class who has gone on a whale watch. Ask him or her to describe the experience and share pictures.

What a Whale!

Use this pattern to make a whale puppet. Just follow the directions on the next page.

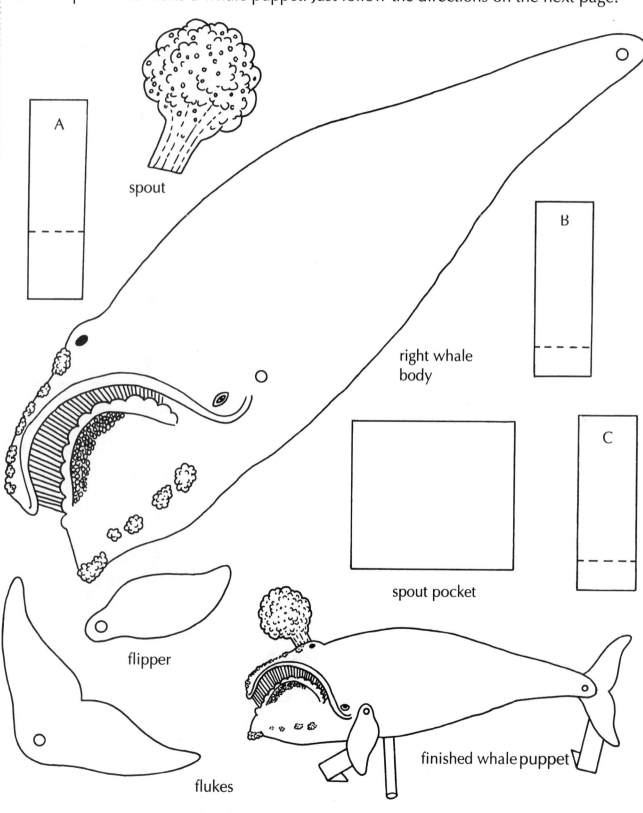

A

spout

right whale
body

B

C

spout pocket

flipper

flukes

finished whale puppet

What a Whale! (cont.)

 What you need:
whale pattern (page 81) • glue • oak tag
• crayons • scissors • tape • straw

 What you do:

1. Glue the page with the pattern on it to the oak tag.

2. Color the whale.

3. Cut out the whale and other parts of the puppet.

4. Tape tab A to the back of the whale's spout, as shown.

5. Tape the spout pocket to the back of the whale's body, as shown. The pocket should be just under the small, dark circle at the top of the whale's body near the fin. (This circle is the whale's blowhole.) Tape only the sides of the pocket. Do not tape the top or bottom.

6. Slip the spout into the top of the pocket you just made.

7. Pull the tab out from the bottom of the pocket and fold tab A back on the dashed line.

8. Glue a straw to the back of the whale shape. Attach flukes and flipper with brackets or glue. Hold the puppet at the bottom of the straw.

9. To make your whale spout, push the tab up.

1.

page

glue

oak tag

4.

spout

tab A

dot on
reverse side

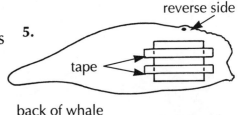

5.

tape

back of whale

spout pocket

6.

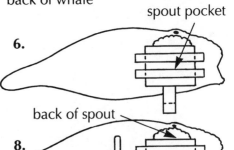

back of spout

8.

folded tab

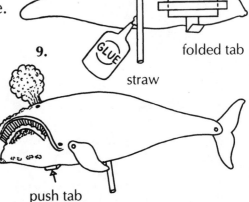

9.

GLUE

straw

push tab

Dear Mr. Blueberry

by Simon James (Margaret K. McElderry Books, 1991, F).

Emily finds a whale in her pond. She writes to her teacher, Mr. Blueberry, for help on how to care for the whale. Mr. Blueberry tells her in each letter that she cannot have found a whale, and offers a multitude of reasons why. The whale leaves after Emily reads to him about whales migrating. Then Emily writes Mr. Blueberry to say that she saw Arthur, the whale, at the beach, and we are left to guess what Mr. Blueberry might reply.

Activities:

1. Whales' eyes are coated with oil to protect them from the salty sea water. Provide each child with two 5 cm squares of brown paper bag. One of the squares should be coated with cooking oil. (Coat these in advance and store them in a plastic bag until needed.) Children will also need water and droppers.

   Initiate a discussion on how our eyes feel when we go swimming. Ask: Have your eyes ever burned from the chlorine in pool water? Have you ever gone swimming in the ocean? How does the salt affect your eyes? Do you suppose whales eyes burn from being in salt water all the time? What might protect a whale's eye? Accept all reasonable answers, then explain that you are going to do a demonstration that will show the special adaptation whales have that enables them to open their eyes underwater without having them burn. Direct the children to examine the two pieces of brown bag. Ask: How are they different? The same? What do you think might happen if a drop of water were placed on each?

   Distribute one oiled and one plain piece of paper to each child. Have the children place the drops of water on the papers, and then describe the shape each takes. Put the papers away on a flat surface for five minutes and examine them again. Ask if the shapes have changed. Have the children describe the shape of each drop. Ask: What do you think helps the one drop remain a drop? Do you think it might last until the end of the day? (Yes, unless it evaporates, as the oil will prevent it from being absorbed by the paper.) Which piece of paper might be more like a whale's eye? Like our eye? Do you think the experiment with the oil and the paper could help you understand how whales can keep their eyes open in salty water without having them sting?

2. Whales have a keen sense of hearing. Sound waves strike the bones in the whale's head and travel to its ears. Ask the children to strike the prongs of a metal dinner fork against a hard surface. Have them place the handle of the fork in their mouth and gently bite on the handle with their teeth.

   Pose the questions: Is the sound different when the fork is in your mouth from when it is struck? How would you describe the sound? Have you ever felt or heard a sound like this before? The vibrations pass from your teeth through the jawbone to the ears.

3. Ask the children if it is easy to recognize their mother or father calling them for dinner. Set up a sound matching game. Place small objects such as paper clips, cereal, beans, rice, and other kinds of seeds into opaque containers, such as margarine tubs, making a pair of each. Cover and distribute to the class. Have students shake the containers and try to match each one with its mate. Ask: If you were a whale, would it be easy to find your family by their "voices" among all the sounds of the ocean?

4. Use the reproducible worksheet on page 85 to help students think about whale facts.

Curriculum Crossovers:

1. Display a wide range of sizes and shapes of empty milk containers. Have the children identify the one most frequently purchased in their homes. Discuss the volume it holds. Ask the children how many glasses of milk they drink a day. Using water, measure the average number of glasses consumed into an empty container. Tell the children that a baby whale may consume 50 gallons a day. A good bulletin board display would be a child with a quart, or 4 glasses of milk, and a baby whale with 50 gallons, or 800 glasses of milk.

2. Use a large wall map to identify the location of Nantucket.

If your community is on an island, ask the children how they think island living differs from living on the mainland. Have any children lived on the mainland? Which do they prefer? Why? How might the children's lives change if they moved to the mainland?

If your community is on the mainland, ask the children if they have ever been to an island. How did they get there? How was life on the island different from where they live? Read a book to the children about island living. Afterward, have students compare life on the island with where they live.

Take a Trip:

1. Take a trip to an aquarium or a sea world (or borrow a video of sea mammals performing). Ask the children to move like the animals they see.

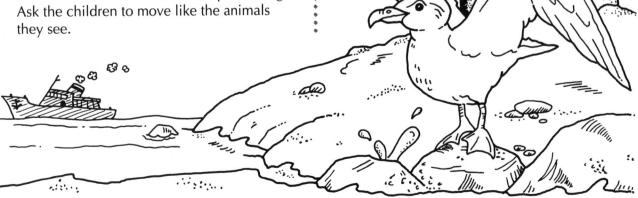

Name _____

Is That a Fact?

Some of the strips at the bottom of this page tell facts about whales. Other strips tell things that are not facts. Paste each strip in the correct column below. If you're not sure something is a fact, use books to help you.

Fact	Not a Fact

whales migrate	whales live in the salty ocean
whales stay in one place all year round	whales eat tiny shrimp-like creatures called krill
whales lose their way in the ocean	whales eat corn flakes and bread crumbs
whales use sounds to find their way in the ocean	whales "talk" to each other by making different sounds
whales live in fresh water ponds	whales don't make any sounds

More About Whales: Find out two more facts about whales. Add them to the "fact" column above.

85

Dinosaurs

* *LITTLE GRUNT AND THE BIG EGG: A PREHISTORIC FAIRY TALE* by Tomie dePaola.

* *THE LITTLEST DINOSAURS* by Bernard Most.

Background:

1. Since humans and dinosaurs did not live at the same time, no one has ever seen a dinosaur. Our knowledge comes from making inferences from fossils.

2. A fossil is the hard part of an animal that changed to rock or a trace of plant or animal life left in rock.

3. Dinosaurs ranged from the size of a chicken to almost as large as a blue whale.

4. Dinosaurs are usually classed as reptiles, but some scientists believe dinosaurs were warm-blooded.

5. Dinosaurs dominated the land. Living at the same time were sea-dwelling reptiles and flying reptiles that are not classed as dinosaurs.

Literature Tie-Ins:

* **DINOSAUR BONES** by Aliki (Thomas Crowell, 1988, NF)

* **DINOSAURS** by Gail Gibbons (Holiday House, 1987, NF)

* **LIVING WITH DINOSAURS** by Patricia Lauber (Bradbury Press, 1991, NF)

* **THE NEWS ABOUT DINOSAURS** by Patricia Lauber (Bradbury Press, 1989, NF)

* **NEW QUESTIONS AND ANSWERS ABOUT DINOSAURS** by Seymour Simon (William Morrow, 1990, NF)

* **WHAT HAPPENED TO PATRICK'S DINOSAURS?** by Carol Carrick (Clarion Books, 1986, F)

Little Grunt and the Big Egg:
A Prehistoric Fairy Tale

by Tomie dePaola (Holiday House, 1990, F).

The story of the trials and tribulations of Little Grunt with his pet George/Georgina. Little Grunt is very sad when he is forced to give up his pet. He remains sad until George saves the entire Grunt tribe from a volcanic eruption.

Activities:

1. Since this book is a fairy tale that imagines people and dinosaurs living at the same time, start by asking the children if they can tell a fairy tale. You may want to read a fairy tale to them and discuss which of the events in the story could have taken place and which would have been impossible. Ask them how fairy tales are different from other fiction books. Point out that the author of *Little Grunt and the Big Egg* knows that dinosaurs and people did not live at the same time but that it makes a good fairy tale.

2. To help children understand when dinosaurs lived in relation to other events in the Earth's history, create a geological time scale. Use 100 paper links to form a chain representing the Earth's 4.5 billion years, with each link representing 45 million years of time. For each new creature, add an extra piece of chain hanging down to which the children can attach a picture of the new organism. When dinosaurs die out, there can be the picture of a dinosaur with a line through it.

Chain Piece	New Development	
	Precambrian Era	
1	[Formation of Earth]	4.5 BYA
29	First single cell life	3.2 BYA
88	First shellfish	570 MYA
	Paleozoic Era	
90	First fish	470 MYA
92	First amphibians	345 MYA
93	Coal formed	300 MYA
94	First reptiles	275 MYA
95	First dinosaurs	225 MYA
	Mesozoic Era	
98	Last dinosaurs and first bird	65 MYA
	Cenozoic Era	
99	(at end) First humans	3 MYA
100	Today	

- BYA = Billion years ago
- MYA = Million years ago

Modern human beings have existed fewer than 40,000 years.

3. Have students make a fossil model. Use a disposable container, such as a cut milk carton, to mix plaster of Paris. Place one cup of plaster in the container and gradually add water, according to directions. Keep stirring the plaster until it is like thick pea soup. Pour the plaster into a cut milk carton or paper cup. Press a shell or other item that has been well covered with petroleum jelly into the plaster and allow it to harden. The shell should come out with a little prying.

4. Have the class brainstorm ways to move a large egg without breaking it. Then have students bring to class something they could use to cradle an egg for a day, and provide each student with a hard boiled egg. The students will be responsible for protecting their eggs from breakage for the rest of the school day. Have a sharing session for children to express how it felt taking care of an egg for a day.

5. Purchase chicken eggs of different sizes and compare their length. The smallest dinosaur, a compsognathus, was about the size of a chicken. Since dinosaurs came in many sizes, dinosaur eggs also have been found in different sizes, from one to ten inches in length. Some are egg-shaped and others are round. Use potatoes of appropriate sizes to represent the sizes and shapes of dinosaur eggs. Have children construct a nest of adequate size to hold the eggs and young dinosaurs. Many dinosaurs laid 12-16 eggs.

6. Use the reproducible pattern on page 88 to have students put together their own dinosaur puzzles.

Curriculum Crossovers:

1. Use a world map to locate volcanoes. Discuss the destruction that can occur when a volcano erupts.

Take a Trip/Invite a Guest:

1. Take a trip to a dinosaur exhibit.

2. Invite an amateur paleontologist to the class to share his or her experiences collecting fossils.

A Dinosaur Puzzle

How do scientists find out about dinosaurs? They study their bones. Putting together a dinosaur's bones is a lot like putting together a puzzle.

Cut out the pieces of this puzzle along the black lines. Then try to put the pieces back together.

Challenge: Turn the pieces of the puzzle over so that you can't see the parts of the picture. Try to put the pieces together without the picture to help you.

The Littlest Dinosaurs

by Bernard Most (Harcourt Brace Jovanovich, 1989, F).

While most people think of dinosaurs as huge creatures, this book takes a light-hearted look at some of the smaller dinosaurs.

Activities:

1. With the help of the children, measure the lengths of some of the biggest dinosaurs. You can do this on the playground. Have students stand in a row with outstretched arms. How many children long is the dinosaur? Have you seen other things that are as big as a dinosaur? What might you do if one came into the school yard?

2. Organize a scavenger hunt. Help children cut string equal to the length of small dinosaurs. With the help of the string, try to find things that are close to, or the same size as, some of the smaller dinosaurs. Have you seen animals that are about this size? How would you feel if one came into the school yard?

3. To simulate the reconstruction of a dinosaur from bones, divide the class into groups and provide each with plastic animal bones or dinosaur puzzles (obtainable from toy stores). Then have the groups reconstruct the animals.

4. Much of what paleontologists know about dinosaurs is based upon inference. To give the students practice in inference, first construct containers from opaque paper cups, preferably hot cups. There should be enough cups for each group of students to have five.

   a. In each cup, make 4 evenly spaced cuts from the rim toward the bottom. The cuts should be about the length of the cup opening.

   b. In each of five cups place a different small object, such as: a die, small block, eraser, marble, penny, or paper clip. Repeat, using identical objects, for each set of cups.

   c. Fold over the cut sections to close the opening and seal them with masking tape. Make sure there are no gaps that allow the object to be seen.

   d. Label each set of cups A-E and distribute them to the groups.

   e. Tell the children they are to practice being scientists and gather information about a "new discovery" in each cup without opening the cup.

   f. Tell them their data is to be descriptive. For example: it rolls, it seems smooth, it has sides, it makes a noise like metal. They are not to say what they think it is. Discourage guessing.

   g. After they have gathered information, display one item at a time and discuss with children whether their inferences were accurate.

Curriculum Crossovers:

1. Using the reproducible worksheet on page 90, students may be able to calculate the relation between their own height and the length of the three dinosaurs shown. How many times longer than the students' height is each dinosaur?

2. Write about what it would be like to live with dinosaurs.

3. Using many descriptive words, tell about an item you have found without naming the item. Other students can then try to guess what it is.

Take a Trip/Invite a Guest:

1. If there is a fossil bed in the locality, take the class there to locate fossils.

Sizing Up Dinosaurs

Some dinosaurs were smaller than you are! The chart below shows how tall some of the smaller dinosaurs were. Use a ruler to find things in your classroom that are the same size or close to the same size as each dinosaur. (For example, a dinosaur might be one notebook tall, two pencils tall, and so on.)

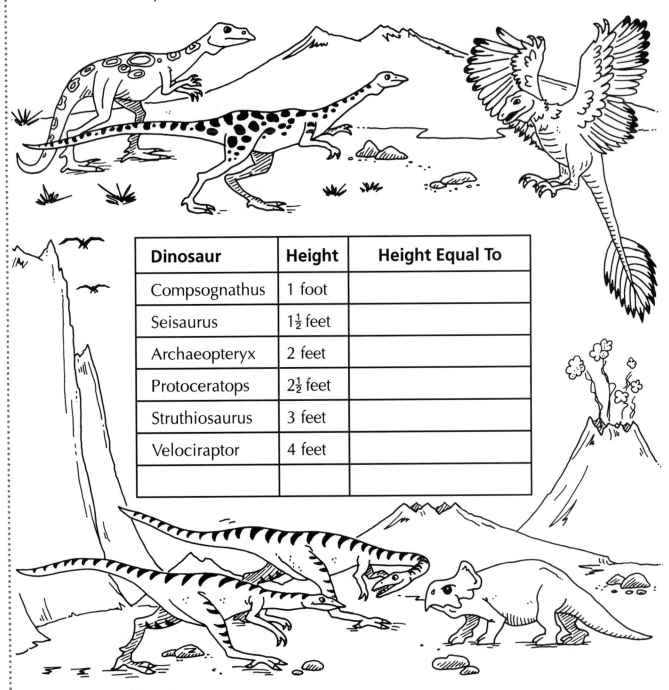

Dinosaur	Height	Height Equal To
Compsognathus	1 foot	
Seisaurus	$1\frac{1}{2}$ feet	
Archaeopteryx	2 feet	
Protoceratops	$2\frac{1}{2}$ feet	
Struthiosaurus	3 feet	
Velociraptor	4 feet	

One More 'Saur: Use books to find one more small dinosaur. Add it to the chart. Compare its height to something in your classroom.

- *PRINCE WILLIAM* by Gloria Rand.

- *THE WARTVILLE WIZARD* by Don Madden.

Background:

1. Oil does not mix with water.

2. Oil coats things it comes into contact with and is difficult to remove.

4. All living things need clean air and water. Oil spills can hurt or kill them.

6. Though litter is not as harmful as oil, it is ugly and costs money to clean up.

7. Conservation of natural resources is necessary if they are to remain available for future generations.

Literature Tie-Ins:

- **COME BACK SALMON** by Molly Cone (Sierra Club Books, 1992, NF)

- **THE HOUSE THAT JACK BUILT** by Ruth Brown (Dutton Children's Books, 1991, F) *Just a Dream* by Chris Van Allsburgh (Houghton Mifflin, 1990, F)

- **MICHAEL BIRD-BOY** by Tomie dePaola (Simon and Schuster Books, 1975, F)

 Prince William

Prince William by Gloria Rand (Henry Holt, 1992, F.).

A child's story of the oil spill cleanup at Prince William Sound. It tells of the rescue and saving of a seal which the child names Prince William.

Activities

1. Provide the children with feathers (obtainable from craft shops or early childhood product suppliers), water, cooking oil, containers, and droppers. Ask them what they think will happen if they place a drop of oil and a drop of water onto a feather. Direct them to place these drops on different parts of the feather. Ask the children to describe what happens.

Ask: What do you think will happen if you submerge one feather in water and another in oil? Have the children submerge a feather in a container of each. Do the feathers behave similarly? Leave them in the water and oil for a week and observe any changes that may occur. Remove them and place them on newspaper for another week. Have the children observe the two feathers. What has happened to them? Based on their observations, help students complete the reproducible data sheet on page 93. Ask: Is oil bad for birds?

2. Simulate an oil spill for the class. Place a clear Pyrex dish of water on an overhead projector. Add drops of cooking oil to the water, keeping count of the number of drops. Have the children observe the results. How do the oil drops behave on the water? Stir the oil and water and observe. Does oil stick to the sides of the dish?

Repeat with thicker types of oil, such as lubricating oil and motor oil. Use the same number of drops for each as for the cooking oil, and wash the dish between samples. How do each of these compare with the cooking oil? Which kind of oil is most like the crude oil of ocean spills? Also compare the ease or difficulty of cleaning the container after each sample.

3. Create an ocean in a bottle. Clean and remove labels from a one-liter plastic soda bottle. If the label is hard to remove, use cooking oil on a paper towel to rub away the glue. If there is an added plastic bottom, loosen and remove it by immersing the container completely in hot water. Now that your container is clear, half-fill it with water. Drop blue food coloring into the water until it is well tinted but not dark. Add cheap mineral oil to the container of water until it is nearly full. Screw the cap on the container as tightly as possible. Turn the container on its side and move the bottom and top alternately up and down. A wave will develop in the bottle. What happens to the oil?

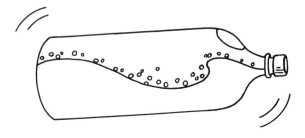

4. Have children apply their newly gained knowledge about oil spills to the art of dyeing eggs. In advance, hard boil enough eggs for each student or group to have one. Make dyes by dropping two different hues of food coloring into containers of warm water. Add one teaspoon of white vinegar to half a cup of dye. Add cooking oil to cover half the surface of the dye mixture. Provide students with a wire holder or spoon to place eggs into the dye mixture. Remove the egg when the proper amount of color is realized. Have the children describe what happened to the egg. Use a paper towel to clean the oil off the egg before placing it in the second color.

When the egg is removed again, have the class describe what has occurred. There will be three colors from the two dyes. The second color will appear not only on areas left untouched by the first color but also on dyed areas, forming a third color.

Curriculum Crossovers:

1. Have students use a map of North America to locate both the spot where they live and Prince William Sound. What states or bodies of water would they pass through to reach the sound?

2. Collect pictures or headlines of oil spills. The reproducible map can be used to identify where those around the United States occurred.

Invite a Guest:

1. Invite a speaker from the Environmental Protection Agency to discuss what the agency does when an oil spill is reported.

2. Invite a bird rehabilitator to the class to discuss how birds are cleaned after they encounter an oil spill.

 # Oil Spill Chart

Use this chart to record your observations of the feather in water and the feather in oil. Use pictures and words to tell what you see.

	Feather in Water	Feather in Oil
After a few minutes in the liquid		
After one week in the liquid		
After one week drying on newspaper		

1. Look at the two feathers after they have dried. How are they different? _____

_____

2. Why do you think oil is bad for birds? _____

_____

The Wartville Wizard

by Don Madden (Macmillan, 1986, F).

An old man who spends his days picking up other people's litter suddenly realizes that he has power over the litter. The people of the town are overwhelmed with their own trash.

Activities:

1. Have the class collect playground litter in large plastic garbage bags. Ask them to sort it into containers for: paper, aluminum, cans, glass, plastic, and tin cans. Have them count the items and determine which group is discarded most often. Children can further explore which group weighs most by using a bathroom scale.

2. Encourage the class to separate the classroom trash and to compare it with the playground trash. Questions to ponder: Does each kind of trash contain the same groups? Do the groups contain the same relative amounts in both kinds of trash? Why might there be differences?

3. Using the reproducible worksheet on page 95, have the children trace each item to its source. Ask: which of the items most commonly become trash?

4. As a followup, have children trace a particular piece of playground litter back to its source.

Curriculum Crossovers:

1. Have each child select one item of playground litter and write about how it got there.

2. Hold a contest to create a poster that would discourage littering.

3. Find out where your town garbage is taken. Locate it on a map. How many miles does the garbage travel to the landfill area?

Take a Trip/Invite a Guest:

1. Take a trip to the landfill area. Count the number of trucks arriving per minute. Ask: Would you like to live near a landfill area? How could we decrease the amount of garbage that arrives here each day?

2. Invite a local person who is involved with efforts to keep highways litter-free and beautiful to talk to the class.

Name

Straight From the Source

Many of the things people use and throw away are made from the Earth's *natural resources*. Trees and oil are two of the Earth's natural resources. Look at the pictures below. Draw a line to match each item with the natural resource used to make it.

Grocery Bag

Book

Toothpick

Ruler

Telephone Pole

Ink

Crayons

Tennis Ball

Frisbee

Toothbrush

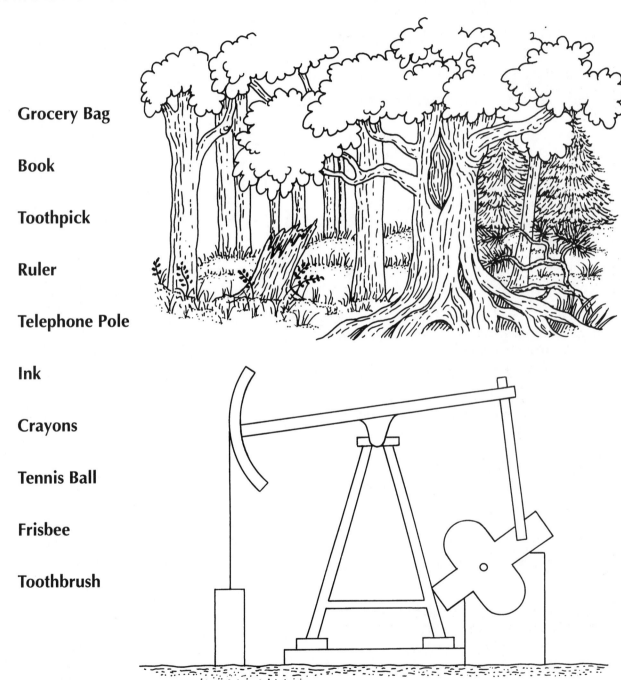

Resource Research: Find out two more things that are made from each of these resources. Add them to the page.

Reduce, Reuse, and Recycle

- *DINOSAURS TO THE RESCUE!* by Laurie Krasny Brown and Marc Brown.

- *STAY AWAY FROM THE JUNKYARD* by Tricia Tusa.

Background:

1. People can reduce the use of natural resources by: substituting reusable items for disposable ones; turning off lights, TV, radio, and computer when not in use; using less water; using paper completely before recycling; buying items packaged in biodegradable materials whenever possible.

2. Reusing items whenever possible, and repairing rather than replacing helps save natural resources and reduces the amount of trash that ends up in landfills and incinerators. If adults model the ideal conservation lifestyle by regularly reducing, reusing, and recycling, children will learn the three r's of conservation by following their example.

3. Recycling saves natural resources and causes less air and water pollution. Recycling and composting save landfill space.

Literature Tie-Ins:

- **THE GREAT TRASH BASH** by Loreen Leedy (Holiday, 1991, F)

- **JET BLACK PICKUP TRUCK** by Patricia Lakin, illustrated by Rosekrans Hoffman (Orchard, 1990, F)

- **RECYCLE! A HANDBOOK FOR KIDS** by Gail Gibbons (Little Brown, 1992, NF)

- **RECYCLING** by Tony Hare (Gloucester; distributed by Watts, 1991, NF)

- **WHAT WE CAN DO ABOUT LITTER** by Donna Bailey (Watts, 1991, NF)

 ## Dinosaurs to the Rescue!

by Laurie Krasny Brown and Marc Brown (Little, Brown, 1992, F).

Slobosaurus provides an example of what *not* to do to save planet Earth. The importance of reducing our use of water, electricity, cars, water, paper, and plastic is stressed. Reasons for reusing and recycling are given on a child's level. The book ends with ways each of us can do our share to give back to planet Earth.

Activities:

1. Have the class create labeled recycling containers in which to collect and sort daily trash. Before it is taken from the classroom, generate a discussion with the following question: Are there items that would not have been in the trash if we had used an alternative? Encourage the class to think about what they regularly use and discard and come up with possible alternatives for those items.

2. Experiment to discover which items readily decompose by setting up a mini-composting experiment. Provide each student with a small, heavy-duty, sealable plastic bag containing organically rich garden soil. (This should come from a back yard or compost pile. Commercial soil is too sterile.) Brainstorm which items belong in a compost pile rather than in the landfill. Ask the children which they think will decompose the fastest. Have each child bring in two or more non-meat food scraps or garden waste to put into their bag of soil. Add a tablespoon of water. Using the reproducible worksheet (page 98), keep a log of how the compost changes during the following weeks.

3. Collect data on water use at the classroom drinking fountain.

First determine the average length of a drink. To do this, select six students and time them as they take an average drink. Delete the times of the students who took the longest and shortest drinks and average the remaining four drinks. Collect and measure the amount of water that runs for the average length of a drink.

Provide a log in which students record the number of times daily they take a drink. Multiply the volume of water in an average drink times the number of drinks the class takes in a day to determine how much water is used for drinking.

Questions to ponder: If water had to be carried a long distance before drinking, would you use as much? Would less water be used if there were cups instead of a fountain? (If so, what about the resources used to make the cups and the need to dispose of them?) How else could less water be used for each drink?

4. As a class, plan a luncheon with conservation of natural resources as the focus. Try to avoid disposable plates, utensils, cups, excess packaging, wasted food, and so on.

Curriculum Crossovers:

1. Have the class weigh the daily waste from the classroom and keep a cumulative record of the amount.

2. Have the class devise slogans encouraging conservation of specific resources.

Take a Trip/Invite a Guest:

1. Take a trip to a water purification plant. Find out how water is purified and what it costs to make water safe to drink.

2. Invite an older person (perhaps a grandparent) who repairs things, such as a plumber, a

tailor, or a clock repair person, to tell the class how repairing things has changed over the years. Also ask him or her to predict how repairing will change by the time the class graduates from high school.

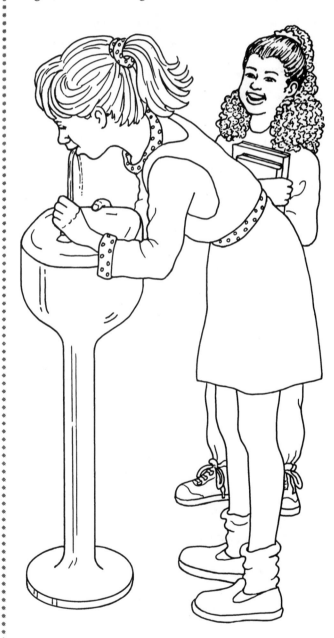

Name

 # Rot On!

Use this chart to record what happens to the trash in your compost pile. Use words and pictures to record your observations. Then answer the questions. (If you touch any trash in your compost pile, be sure to wear rubber gloves.)

Type of Trash	One Week	Two Weeks	Three Weeks	Four Weeks

1. Which piece of trash changed the most after four weeks? _____

2. Which piece of trash changed the least? _____

_____

3. Many people are worried that we are running out of places to put all the garbage we

make. How can compost piles help solve our garbage problems? _____

_____

_____

Stay Away From the Junkyard

by Tricia Tusa (Macmillan, 1988, F).

When Theo goes to visit her aunt, everyone except her aunt warns her to stay away from the junkyard. Theo finds it a wonderful place with all sorts of marvelous playthings, including a pig.

Activities:

1. Distribute the reproducible data sheet (page 100) and request the help of parents to work with their children in collecting and recording data on items their family discards. Have the children bring their charts to class. Summarize the data on a graph. Questions to ponder: Are there some things we use a lot of? What do we as a class use little of? Is there something we should use less of?

2. Start a discussion on packaging by displaying an overpackaged item of interest to a child. (Such items are easily found near the checkout counters of discount stores.) Ask: Why do products need packaging? List the children's responses on the board. Have children bring the packaging to class when a new item is purchased at home. Examine the packaging. Ask: What is it made of? Will it decay if composted? Does the packaging take up space? Why was the packaging necessary? Are there other ways to package the item?

3. Brainstorm as many ways as possible to reuse a commonly discarded item such as an empty toilet paper tube. Draw diagrams showing

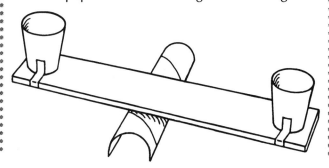

how to use a toilet paper tube to make a balance.

4. Hold a creative "new from old" contest. Ask each child to carefully examine items that are discarded in their homes and to select one or more to reuse in a creative way.

Curriculum Crossovers:

1. Write the "life story" of the item(s) you used for the "new from old" contest.

2. Organize a treasure hunt for objects made from recycled materials. Assign it as homework. Students will search their homes to find items with the recycling emblem (now displayed on many household items, like books, cans, boxes and bottles.) Students can bring in the emblem only, or the entire, cleaned item. Chart or graph results of the treasure hunt.

Take a Trip:

Take a trip to a thrift shop to see all the material for sale. Focus on the question: Why is it important to reuse items that are still in good condition?

Be a Garbologist!

People who study garbage are called Garbologists! They can tell a lot about people just by studying the garbage we throw away. A garbage can that contains a lot of disposable diapers, for example, probably belongs to a family that has a new baby. Garbage that contains a lot of G.I. Joe or Barbie wrappers probably belongs to a family with young children. What could a garbologist tell about your family just by studying your trash? Use the chart below and your thinking skills to find out! Count the kinds of garbage listed below every day for one week. Fill in the amounts of each item in the box for that day. At the end of the week, see how much garbage you've counted!

	Mon.	Tues.	Wed.	Thurs.	Fri.	Sat.	Sun.
Glass Bottles							
Aluminum Cans							
Juice Boxes							
Writing Paper							
Newspapers							
Plastic Containers							
Food Wrappers							
Food							
Other							

What do you think your family's garbage tells about you? How do you think this information could help us care for our environment?

The Milky Way Galaxy

- *MY PLACE IN SPACE* by Robin and Sally Hirst.

- *SKY ALL AROUND* by Anna Grossnickle Hines.

- *MARCELLA AND THE MOON* by Laura Jan Coats.

Background:

1. Our sun is one of hundreds of billions of stars in the Milky Way Galaxy.

2. The sun, its nine planets, the asteroids (tiny planets or large chunks of rock), and the planets' moons make up our solar system.

3. Moons revolve around planets and planets revolve around the sun.

4. Earth is the third planet from the sun.

5. Earth rotates on its axis once every 24 hours, giving us a day.

6. Earth revolves around the sun once every 365 1/4 days, giving us a year.

7. Our moon reflects light from the sun. It can often be seen in the sky not only at night but also in the early morning and late afternoon. Approximately once a month it goes through different phases from new moon to full moon and back.

8. Many of the bright stars in our galaxy are visible on a clear night.

9. Constellations are groups of stars that appear to form figures or shapes in the night sky. Eighty-eight constellations have been identified and named.

10. There are many other galaxies besides the Milky Way, but nearly all are too distant to be seen by the naked eye.

Literature Tie-Ins:

- **NORA'S STARS** by Satomi Ichikawa (Philomel Books, 1989, F)

- **GLOW IN THE DARK CONSTELLATIONS** by C.E. Thompson (Grosset & Dunlap, 1989, NF)

- **THE SKY IS FULL OF STARS** by Franklyn M. Branley (Thomas Y. Crowell, 1981, NF)

- **THE PLANETS IN OUR SOLAR SYSTEM** by Franklyn M. Branley (Thomas Y. Crowell, 1981, NF)

- **THE MAGIC SCHOOL BUS LOST INSIDE THE SOLAR SYSTEM.** by Joanna Cole (Scholastic, 1990, F)

- **MOONCAKE** by Frank Asch (Scholastic, 1987, F)

 ## Sky All Around

by Anna Grossnickle Hines (Clarion Books, 1989, F).

A father and daughter observe the night sky.

Activities:

1. Students can make constellation projectors from oatmeal boxes or large paper cups. Distribute copies of the constellation patterns on page 103, and allow students to cut one or more out along the dotted circles. Distribute paper cups or oatmeal boxes and glue, and help students glue one constellation to the bottom of each cup. Use a pencil to poke holes in the dots representing stars. Place a flashlight in the open end of the cup or box and switch it on to project the constellations on a darkened wall.

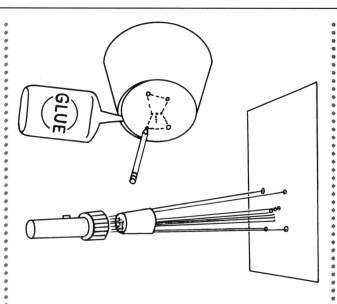

2. Make a mini-planetarium. Obtain a large appliance box from a store. Cut a small opening to act as a door. Children can bring their constellation projectors in and shine the constellations on the ceiling of their planetarium.

Curriculum Crossovers:

1. Using an encyclopedia or other source, compare the legends of Native American and Greek mythology about the constellations.

2. Create a class constellation and write a myth about it.

3. Listen to Jiminy Cricket's song, "When You Wish Upon A Star." Why do you think people "wish upon a star"? What would you wish for on a star?

Take a Trip:

1. Have the children go outside on a clear night with an adult and look at the night sky. Ask the children to sketch what they see. Does the sky seem to change from night to night?

2. Visit a planetarium and learn about the changing constellations during the different seasons.

Star Search

Choose one of the constellation patterns on this page, and cut it out along the dotted lines. Glue it to the bottom of a paper cup or round oatmeal box. Use your pencils to poke holes in the dots representing stars. Place a flashlight in the open end of the cup, and with your classmates, take turns projecting your constellations onto your classroom walls.

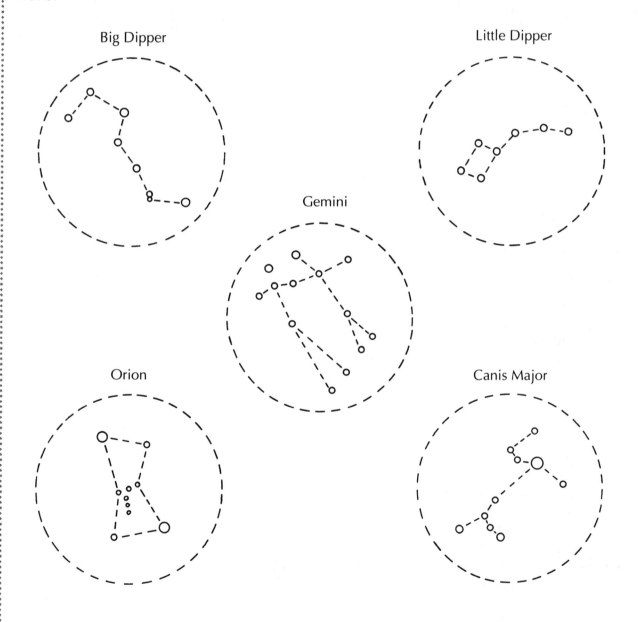

Big Dipper

Little Dipper

Gemini

Orion

Canis Major

More to Do: Many constellations are named after characters in stories from long ago, called "myths." Your teacher can help you find books that will tell you more about myths and constellations.

Marcella and the Moon

by Laura Jan Coats (Macmillan, 1986, F).

Marcella doesn't want to swim with the other ducks. She would rather paint the moon as it goes through its phases.

Activities:

1. Monitor the changing phases of the moon. Have students go out with an adult on key nights to "paint" the moon as it appears in the sky.

2. To demonstrate how the moon shines, darken the room and point a flashlight onto a reflecting object such as a bicycle reflector. The reflector glows only when the light is on. The moon does not give off its own light but simply reflects light from the sun.

3. Cover a styrofoam ball with aluminum foil. Stick a pencil through the ball. Place a lightbulb on a stand that is higher than the children. Children hold the pencil still and rotate their bodies to observe the phases of the moon.

4. Using the reproducible worksheet on page 105, have the children color and create a moon phases flip book.

Curriculum Crossovers:

1. Have the children draw pictures of the same scene by night and by day. They can also draw pictures of what they do at night and during the day.

2. Create a diary of descriptions of what the moon sees from day to day as it looks down on the classroom, the neighborhood, etc.

Take a Trip/Invite a Guest:

1. Find the moon during the daylight hours. The best times are early in the morning just after a full moon, or late in the afternoon just before a full moon. Note: depending on where you live, the brightness of the sunlight could make viewing the moon more difficult at those times.

2. Invite a local amateur astronomer into your class. Ask the astronomer to bring equipment and explain what it is used for and how it works.

The Moon Shapes Up

Name That Moon: Use books to find out what each phase of the moon is called. Write the name of the moon phase under each picture in your flip book.

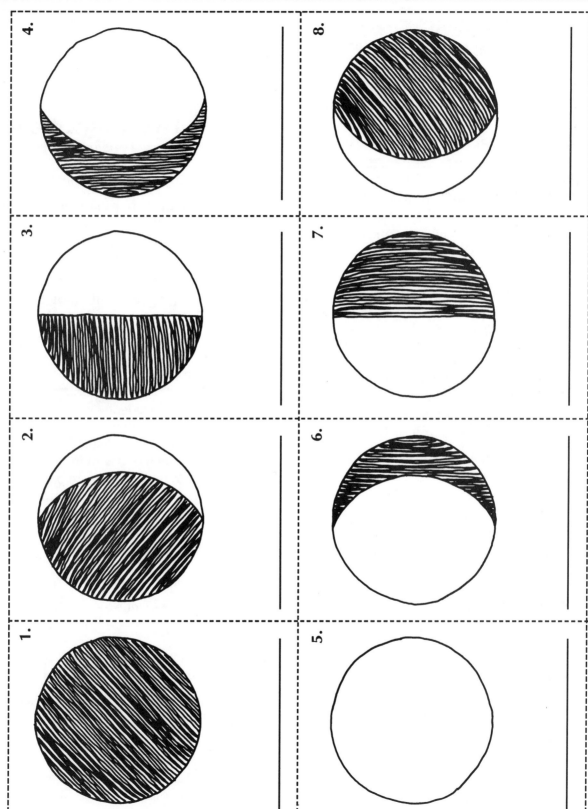

4.

3.

2.

1.

8.

7.

6.

5.

- *THE MAGIC SCHOOL BUS INSIDE THE EARTH* by Joanna Cole.

- *THE SUN, THE WIND, THE RAIN* by Lisa Westberg Peters.

Background:

1. Earth is a solid sphere with an outer covering called a crust, a middle layer called the mantle, and a center called the core. The boundary between the crust and the mantle is called the moho.

2. The inner core and parts of the mantle are so hot that they are molten. The molten rock in the mantle, called magma, causes volcanic eruptions.

3. The crust is made up of rocks, dirt, and water.

4. Rocks may be sedimentary (dirt pressed together), igneous (molten and then hardened), or metamorphic (shaped by both heat and pressure).

5. Dirt can have many textures, weights, smells, colors, and compositions. It consists of broken down rocks and the remains of dead plants and animals.

6. The Earth's surface is constantly changing. Very slowly, new mountains are being formed. At the same time, erosion (the effect of water, ice, and wind) is wearing away the surface.

Literature Tie-Ins:

- **EVERYBODY NEEDS A ROCK** by Byrd Baylor (Aladdin Books, 1974, F)

- **HOW TO DIG A HOLE TO THE OTHER SIDE OF THE WORLD** by Faith McNulty (Harper Collins, 1990, F)

- **THE REAL HOLE** by Beverly Cleary (William Morrow, 1986, F)

- **SYLVESTER AND THE MAGIC PEBBLE** by William Steig (Simon and Schuster Books, 1969, F)

 The Magic School Bus Inside the Earth

by Joanna Cole (Scholastic, 1987, F)

Ms. Frizzle takes her class on a field trip to the center of the Earth, passing through the crust, mantle, outer core, and inner core before coming out through a volcano on the other side. The class learns about sedimentary rocks, containing fossils, that change into metamorphic rocks and then heated igneous rocks.

Activities:

1. Provide each student with a peanut M & M to represent the Earth (you may want to get parents' permission first). Have the students carefully bite the M & M in half and remove one half to look at. (Be sure part of the peanut remains in this half.) Have them describe what they see. Tell them to think of the M & M in the following manner:

Candy shell = Earth's Crust

Sugar coating = Moho

Chocolate = Mantle

Peanut = Core

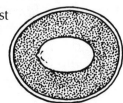

Compare this model with the actual layers of the Earth. How are they alike or different?

2. Using the reproducible activity sheet on page 108, have the children make and label a model of the layers of the Earth.

3. Obtain permission to dig a hole in the school grounds as a whole-class project. Have students compare the layers of soil as they go farther down. At each layer tell the children:

Rub the soil between your fingers. Describe what it feels like. Drop some soil onto paper. How does it sound? Compare the soil's color between layers. Did you find anything that is or was alive? Count the number of stones in each layer and compare their size.

Have students measure the size of the hole and compare it to the size of a familiar item. Ask: What do you think will happen to the hole and to the soil pile when it rains? After a rainstorm, have the students observe and describe the changes.

Before any rain falls, use a balance to compare the mass (weight) of equal amounts of soil samples from the top and bottom of the hole. Allow the samples to dry in the sun and reweigh them. How have they changed?

4. Have pairs of students put an equal amount of topsoil and subsoil into separate funnels. To keep soil from falling through the opening, use an inexpensive coffee filter in the funnel.

Add equal amounts of water to each. (35mm film canisters are all the same size and could be used for measuring the water.) Have students carefully observe and record: Which sample had water come out of the funnel first; which had the greatest amount of water come through the funnel; and any other differences they might observe. Students should discover that topsoil, which contains organic matter, holds water better than subsoil. What will happen if topsoil is lost (by erosion or human activities)?

5. Have students bring in soil samples from home to compare with each others'. Use a magnifier to look at the samples.

Add an equal amount of several different samples to a jar and twice as much water. Cover the jar and shake vigorously. Allow to settle overnight. Ask the students to draw a picture of what they think the contents will look like in the morning. They will probably be surprised to see the distinct layers of sediment.

Curriculum Crossovers:

1. Serve the class "Edible Dirt," using the following recipe. (It looks like dirt but tastes delicious — a reminder to children that observation can depend on all of the senses, not just on vision!)

- 2 lbs. of chocolate sandwich cookies
- 2 boxes instant vanilla pudding mix
- Amount of milk called for on vanilla pudding box
- 1 8-oz. package cream cheese
- 1 cup powdered sugar
- 1 stick (1/2 cup) margarine
- 1 16-oz package of thawed non-dairy topping (such as Cool Whip)
- A small piece of aluminum foil or waxed paper
- A clean flowerpot

Blend the cookies in a food processor until they look like potting soil; set aside. In a bowl, use an electric mixer to blend the pudding, milk, cream cheese, powdered sugar, and margarine. Fold the non-dairy topping into this mixture. Place the foil or waxed paper on the bottom of the clean flowerpot to cover the drainage hole, if there is one. Put some of the cookie crumbs in the bottom of the flower pot and press along the sides. Pour the pudding mixture carefully into the flower pot. Cover with the reserved cookie crumbs. A plastic flower can be planted in the mixture and even a few gummy worms added. Serve and enjoy!

Take a Trip/Invite a Guest:

1. Visit a newly dug house foundation, water or sewer line hole, or road cut. Look at the wall of dirt and compare the color and texture from top to bottom.

A Look Inside

What's inside the Earth? Make a model of the Earth and see for yourself. Here's how.

 What you need:
crayons • scissors • glue • pen or pencil

 What you do:
1. Use the crayons to make each circle a different color.
2. Cut out the circles.
3. Number the circles from smallest to biggest. The smallest circle should be #1. The biggest circle should be number #4.
4. Glue circle #1 to the center of circle #2.
5. Glue these two circles to the center of circle #3.
6. Glue these three circles to the center of circle #4.
7. Read the definitions at the bottom of the page for each of the Earth's layers. Then write the name of each layer on your model of the Earth.

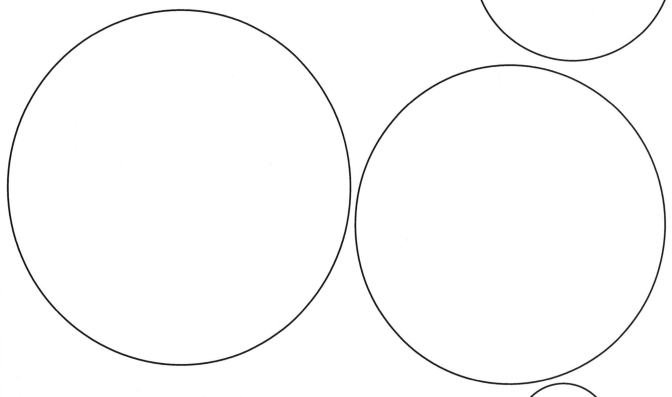

Crust: the solid, outer layer of the Earth
Inner Core: the hot center of the Earth made up of melted rock
Outer Core: the layer of the Earth beneath the mantle
Mantle: a layer of the Earth under the crust and over the core

The Sun, the Wind, the Rain

by Lisa Westberg Peters (Henry Holt, 1988, NF).

Tells two stories on opposite pages. One describes the ongoing changes in Earth's crust, while the other presents a little girl who builds a sand castle and then watches as the tide washes it away. The stories and pictures demonstrate the building up and wearing down of Earth's crust.

Activities:

1. Demonstrate the eroding of a sand castle by water. Pack sand into a plastic glass and dampen it with water to make a mold. Invert the mold at the high end of a pan used for roller painting. Remove the plastic glass, leaving the sand mold. Sprinkle water over the sand and observe what happens. Ask the class to compare this with what happened to the little girl's sand castle.

2. Ask the class if they have ever watched something crumble. What else might crumble under the right circumstances? Provide each student with a baby food jar of warm water and a sugar cube. Ask the students what they think will happen when the cube is placed in the water. Have them place the sugar cube in the water, observe, and describe what happened.

   The gradual breaking apart and dissolving of the cube is a simulation of weathering of the Earth's surface.

3. Have the students use pocket microscopes or magnifiers to compare sand samples.

   You can make sand sample slides out of halved index cards. With a paper punch, make a hole in each halved card. Cover one side of the hole with a piece of scotch tape. Holding a finger over the other side, place the sticky side into a sand sample. Cover the sand with another piece of tape.

   Place the pocket microscope on the index card over the sand and adjust the focus. (With a magnifier, hold it close to the eye and adjust the focus by moving the sand sample to and fro.) Have the children draw what they see in each of the samples. They may need colors for accurate portrayal.

4. Simulate magma turning to lava after it erupts from a volcano. Provide each student with a sealable plastic bag containing some plaster of Paris. Add water to the bag and seal. Have students squish the sealed mixture with their fingers. Ask them to describe the feel. (The plaster will get warm before it turns hard.) Encourage them to make shapes that will form a fake rock when set. When the plaster of Paris dries, the bag can easily be removed.

5. Use the reproducible data sheet on page 110 to help students understand erosion.

Curriculum Crossovers:

1. Have the class write letters to obtain samples of sand or soil from other parts of the country. They can write to relatives and/or exchange local samples with other schools. You can obtain school contacts by writing to the National Science Teachers Association, 1742 Connecticut Ave., Washington D.C., 20009-1171. When samples arrive, locate where they came from on a map. You could attach a small sample to a file card and use a piece of yarn to connect the sample with a map location. Students can keep a log of samples received.

Take a Trip:

After a heavy rainstorm, ask the class to pretend that they are detectives going to look for examples of erosion on the school grounds. Their job will be to determine how the erosion happened and what can be done about it.

Washing Away

Rain can change the way the land looks. Try this experiment and see how.

⬤ **What you need:**
2 index cards • 2 popsicle sticks • tape

⬤ **What you do:**
1. Tape a popsicle stick to the back of each card.
2. Find a grassy spot somewhere outside your school. Push the popsicle stick into the soil. The edge of the card should touch the ground.
3. Find a spot where the soil is bare. Push the second card on the popsicle stick into the ground.
4. Check the cards after a rain. Draw a picture of what each card looks like.

Card in bare soil

Card in grassy soil

5. Which has more mud-splatter marks, the card in the bare soil or the card in the

   grassy soil?_____

6. The mud-splatter marks are a sign that the rain is wearing the soil away. This is called erosion. Based on this activity, how can you help keep the rain from

   wearing soil away? _____

   _____

Resources for Teachers

○ **ANIMAL TRACKS** by Arthur Dorros (Scholastic, 1991, F)

○ **ASTRONOMY** by Jerry DeBruin and Don Murad (Good Apple, 1988, NF)

○ **THE BIG BEAST BOOK** by Jerry Booth (Little, Brown, 1988, NF)

○ **BOOKS KIDS WILL SIT STILL FOR - THE COMPLETE READ-ALOUD GUIDE** by Judy Freeman (R.R. Bowker, 1990, 1992, NF)

○ **BUGS** by Nancy Winslow Parker (Greenwillow, 1987, NF)

○ **CREATURES WITH POCKETS** by Susan Harris (Franklin Watts, 1980, NF)

○ **CRINKLEROOT'S BOOK OF ANIMAL TRACKING** by Jim Arnosky (Bradbury Press, 1987, F)

○ **DINOSAURS: A JOURNEY THROUGH TIME** by Dennis Schatz (Pacific Science Center, 1987, NF)

○ **DISCOVERING SLUGS AND SNAILS** by Jennifer Coldrey (The Bookwright Press, 1987, NF)

○ **GREEN THUMBS UP! STEP INTO SCIENCE** by Barbara Taylor (Random House, 1992, NF)

○ **IN THE LICK OF A FLICK OF A TONGUE** by Linda Hirschmann (Dodd, Mead, 1980, NF)

○ **KEEPING MINIBEASTS: SNAILS AND SLUGS** by Chris Henwood (Franklin Watts, 1988, NF)

○ **SEEDS, JUNIOR SCIENCE** by Terry Jennings (Gloucester Press, 1990, NF)

○ **SOME FEET HAVE NOSES** by Anita Gusta (Lothrop, Lee & Shepard Books, 1983, NF)

○ **SOMEONE SAW A SPIDER: SPIDER FACTS AND FOLKTALES** by Shirley Climo (Thomas Y. Crowell, 1985, NF)

○ **SPIDERS** (Lerner Publications, 1981, NF)

○ **SPIDERS** by Norman Barrett (Franklin Watts, 1989, NF)

○ **TEETH TUSKS AND FANGS** by Roger Dievart (Young Discovery Library, 1991, NF)

○ **THIS BOOK IS ABOUT TIME** by Marilyn Burns (Little, Brown, 1978, NF)

continued

○ **WHERE ARE THE DINOSAURS?** PALEONTOLOGY FOR KIDS by Barbara S. Rountree, Nancy Y. Taylor, and Melisa B. Shuptrine (The Learning Line, 1989, NF)

A field guide series for each classroom is a valuable resource. There are many guides available at different levels of sophistication, including:

○ **Golden Guides, Golden Field Guides** (Golden Press, Western Publishing Co.). The Golden Guides are for beginners.

○ **Peterson's Field Guides** (Houghton Mifflin Co)

○ T**he Audubon Society Field Guides** (Alfred A. Knopf, Inc)

○ **Crinklefoot's Guides**, by Jim Arnosky (Bradbury Press)

Appendix

Care and Maintenance of Land and Water Snails

Cover the bottom of a five gallon aquarium with a 5 cm-deep layer of gravel and add soil sloping up from a depth of 5 cm at one end to 12 cm at the opposite end. Cover the surface of the deeper end with a thin layer of dry leaves and small bark-covered twigs. Keep the container moistened at all times. Cover with a glass plate, leaving enough space for air circulation.

Feed the snails three times a week with a mixture of dry rolled oats, or lettuce or romaine sprinkled with calcium carbonate (available in drugstores in powdered form, or in pet shops as "cuttlefish" or "cuttlebone".) If using the cuttlebone form, simply place in the bottom of the tank.

Do not overcrowd the snails. Remove uneaten food to prevent moldering. Keep snails out of direct sunlight. They should remain active at room temperature. To motivate snails out of estivation (short-term hibernation), mist the terrarium lightly.

Young Children and Temperature Measurement

Young children can and do enjoy measuring things. To insure their success with measuring temperature a few adaptations are helpful.

Color-code thermometers. (Thermometers with some kind of background card are best.) Paint the side of the thermometer by the lower temperatures blue. Paint the side of the thermometer by the higher temperatures red. Paint the in-between temperatures yellow. Students can now record temperatures by color.